Thermodynamics

Thermodynamics

A Rigorous Postulatory Approach

S.H. Chue

University of Malaya
Kuala Lumpur, Malaysia

JOHN WILEY & SONS

Chichester · New York · Brisbane · Toronto

Library of Congress Cataloging in Publication Data:

Chue, S.H
   Thermodynamics : a rigorous postulatory approach.

   Includes index.
   1. Thermodynamics.
I. Title.
QC311.C496   536'.7   76-44878

ISBN 0 471 99455 3 (Cloth)
ISBN 0 471 99461 8 (Paper)

Filmset in India by
Oxford Printcraft India Private Limited, New Delhi & Gandhidham
Printed by The Pitman Press, Bath, Avon

To my parents

'And thou shalt teach them ordinances and laws and shalt
show them the way wherein they must walk'

Exodus XVIII: 20

Contents

Preface

Thermodynamics is the branch of science dealing with our understanding of the properties of matter as they are affected by changes in temperature. It began as an experimental pursuit into the phenomena related to thermal interactions between bodies. However its usefulness was realized by different groups of people at different periods of its development (mainly because each area was then built up from a separate collection of experimental data that seemed to bear little resemblance to other areas. As a result, many textbooks on the subject have been written for each of its various specializations with little cross reference to other fields. For instance, texts on heat engines usually contain little other information beyond that strictly necessary for its exposition, and chemical or chemical engineering texts are primarily devoted to the study of topics in physical chemistry. However, a much wider theoretical basis now exists on the subject whereby each different topic can be neatly accommodated into a body of self-consistent theory. This systematic study of the subject was brought about by Hatsopoulos and Keenan (co-authors of the book entitled 'Principles of General Thermodynamics'). However, their work demands a level of sophistication generally higher than that possessed by undergraduate students.

In the belief that a rigorous development of the subject is of the highest importance in that it not only permits a more elegant exposition of the subject matter but also helps to cut across compartmentalized areas of specialization, thus stimulating interdisciplinary teaching, the author has taken this opportunity to introduce a text at the undergraduate level while retaining the rigour of Hatsopoulos and Keenan's presentation. Such an approach should be particularly beneficial to the physicist as this allows him to see with greater clarity the order of the physical world. In order to achieve the aim of injecting greater rigour into the undergraduate course, the text is based on the laws of thermodynamics rather than on the generalized principles of Hatsopoulos and Keenan. From these laws, the student will be guided to all other topics generally treated in a thermodynamics course by means of formal logic. Though some knowledge of calculus is necessary for the more advanced topics in the text the degree of dexterity required is not very great.

The author has felt free to tidy up some oddities that still persist in the thinking of many thermodynamicists. This should help to make the treatment of the subject more rational than it is at present. In the past, many writers had maintained that the practical temperature scale and the theorists' concept of thermodynamic temperature referred to two different entities; indeed, both scales are accepted even today. Such a contention can be attributed to the fact that the first attempts to measure temperature were made long before any thermodynamic studies. To remove this anomaly, the present discussion on temperature begins with the qualitative definition of Maxwell. This is taken through the zero and first laws of thermodynamics to the second law, wherein it is evident that the logical exposition of thermodynamics does not require any quantitative measures of temperature before it can be so defined. It is hoped that eventually the kelvin alone will be used as the unit of temperature. The layman should in time have no difficulty in realizing that ice forms at about 273 kelvins, a heat wave sets in at about 310 kelvins and that our comfort range lies somewhere below 300 kelvins.

Similarly, since all experimentally observed laws in physical chemistry can be deduced as a consequence of a basic definition of an ideal solution, the treatment of ideal solutions in this text is considered more consistent than that of Hatsopoulos and Keenan.

The subject matter has been arranged in such a way that students following an applied thermodynamics course could proceed onto engineering cycle analysis, without confusion as to the choice of properties to be used in their calculations. To achieve this objective, the topic on properties of pure substances has been split into two parts, with cycle analysis inserted between them. Thus, students in engineering schools can be given a first course in thermodynamics by covering Chapters 1 to 6. More advanced topics on combustion and gas mixtures with an appropriate selection of materials, and more advanced topics on cycle analysis can be given in a second course. This arrangement does not break the continuity of presentation for students in chemical engineering schools as the topics on application are also well placed. Students of chemistry and physics may omit Chapter 6 or gloss over it with a few examples to gain an idea of the type of analysis that can be performed with the first and second laws.

The last two chapters present some topics of special interest. In particular some alternative schemes for energy conversion to those given in Chapter 6 have been included in greater detail than generally found in texts on thermodynamics, in the hope that it will encourage students to further study of the subject. It is impossible to give adequate acknowledgements to all those who have contributed either directly or indirectly to this book. However, much is owed to Professors George N. Hatsopoulos and Joseph H. Keenan, whose book inspired the writing of this text. The author is particularly indebted to them for their kind permission to produce in the abridged form their discussions on the *Phase Rule* and on the *effect of chemical equilibrium on the number of components* of a multi-component system. Their permission to use two of their problems, *Problems 11.5 and 12.1*, is also gratefully acknowledged.

S.H. Chue

Kuala Lumpur
May, 1976

CHAPTER ONE

Fundamental Concepts

1.1 INTRODUCTION

Thermodynamics is the branch of science dealing with our understanding of
matter as it is affected by changes in temperature. To date two approaches
have been found successful in its description. These are;

(i) postulatory (or classical);
(ii) statistical.

The postulatory approach was originally developed primarily for the study of
heat engines. However, its scope has since been extended to many other processes. It is not concerned with the form of matter but seeks to relate the observed
physical properties of matter from using certain basic postulates. In contrast, the
statistical approach is founded on the atomistic behaviour of matter and seeks
through molecular dynamics to justify the postulates and other phenomenological laws taken for granted in the postulatory approach. We are not concerned here as to which of these two approaches is the more fundamental or more
exact, or which approximates to the other, but since the mathematics required
in the statistical approach is much more sophisticated, this will be left to more
advanced courses.

Though historically classical thermodynamics began with experiments on
a macroscopic scale, the results of which were then unified into a self-consistent
theory, thermodynamic principles can best be explained from a purely logical
viewpoint. Using this approach, we shall first familiarize ourselves with the
basic definitions and fundamental concepts that are essential to our understanding of the subject matter to be presented in later chapters.

1.2 BASIC DEFINITIONS

To start, a *system* may be defined as an isolated region in space enclosed by well
defined surfaces across which no mass can pass. The entire region which includes
everything external to the system is called its *surroundings*. Surfaces separating
the system from its surroundings are known as *boundaries*. Ideally, boundaries
are mathematical surfaces which we endow with various ideal physical characteristics, such as rigidity and impermeability. If a system has been isolated

from its surroundings, all changes occurring in the system are considered to be independent of all changes that take place in the surroundings; hence if two systems are not isolated from each other, then changes in one will affect changes in the other. Since no mass can pass from one into the other, we invariably conclude that such changes are brought about by means of *boundary interactions* only.

As thermodynamics primarily deals with changes in systems, measures for identifying these changes are necessary. These measures are generally based on certain characteristics that can describe the condition of the system under investigation, and must be found by experience since many physical character-istics are irrelevant from the thermodynamic viewpoint, for examples, colour and shape. Any suitable characteristic whose value depends on the condition of a system and which is relevant to our thermodynamic investigation is known as a thermodynamic *property*.

There are two fundamentally different ways in which we can classify properties. From the standpoint of measurability, the properties may be termed *primitive* or *derived*. By primitive property, we mean any property that can be specified by an operation on the system that does not effect any noticeable change on the system. Any property that is not a primitive property is known as a derived property. Alternatively, from the computational viewpoint, the properties may be *intensive* or *extensive*. The former are essentially local in character and include such quantities as pressure, temperature, density and electric field. Mathematically, an intensive property is defined using the concept of a limit at a point, so that such properties are independent of the size of the system. Note that the mathematical definition deals with a hypothetical contin-uous substance, i.e., the continuum, and for the definitions to be valid, the 'point' taken for the limit should remain several orders larger than the molecular mean free path or distance so that coarseness in the fine structure of matter does not emerge. Properties of the extensive class in a sense convey a measure of the size or extent of the system, and include such quantities as mass, volume and internal energy. These quantities are generally proportional to the mass of the system if other conditions are kept constant. Thus, using the *additive rule*, the value of an extensive property is equal to the sum of its values for all parts of the system into which it may be subdivided.

It is often convenient to refer to extensive properties in terms of their values *per unit mass* of the system. These are known as *specific* properties. It has been the usual practice to denote extensive properties by capital letters and their derived specific properties by the corresponding small letters. For instance, we denote the volume of a system by V and the specific volume by v. Note that specific properties are also local in character and hence have often been referred to as intensive properties in many text books.

To continue, the *state* of a system is defined as that condition of the system which is capable of being completely specified by all its primitive properties. The entire series of states that a system undergoes during a change of state is called the *path* of the change, while the complete description of the change of a system including the end states, the path and the boundary interactions constitutes a *process*. A process whose end states are identical is referred to as a *cycle*. It should be noted here that classical thermodynamics requires that processes should be carried out infinitely slowly, as shown in Section 1.5. The actual rate of a process is the concern of irreversible thermodynamics which will be examined briefly towards the end of this book.

As a consequence of the above definitions, we arrive at the following useful results, which we shall refer to as corollaries to the basic definitions.

Corollary 1 A change of state is fully described by means of the initial and final values of all the primitive properties of the system. A change occurs when at least one of its primitive properties changes value.

Corollary 2 A process is required for the determination of a derived property.

Corollary 3 The change in value of a property is fixed by the end states of a system undergoing a change of state and is independent of the path.

Corollary 4 Any quantity which is fixed by the end states of a process is a property of a system.

Corollary 5 When a system goes through a cycle, the change in value of any property is zero.

Corollary 6 Any quantity whose change in a cycle is zero is a property of a system.

These corollaries will be referred to many times in later discussions and students are therefore advised to become thoroughly familiar with them.

1.3 WORK AND HEAT

In mechanics, *work* is defined as the product of the displacement of a force and the component of the force in the direction of the displacement. Thus

$$W = F_x \cdot x \tag{1.1}$$

where F_x denotes the component of F in the x direction and x the displacement.

If F_x depends on x, then eqn. (1.1) can be written in the form of an integral

$$W = \int F_x \mathrm{d}x \tag{1.2}$$

This is a useful definition as it enables us to evaluate the work done by a force.

However, in view of the fact that we are dealing primarily with interactions between systems in thermodynamics, the above definition is not directly applicable. If we examine this definition more closely, we note that even in the mechanical sense, work is done only when the force moves, no work being done when the force remains stationary, i.e., the definition of work involves a process. From this consideration, we conclude that work cannot be a property of a system. Rather, it is something that the system experiences while it changes from one state to another. We are now in a position to extend the mechanical definition of work to the thermodynamic sense; thus

4

Work is an interaction between two systems such that what happens in each system at the boundary could be repeated with the change in level of a weight as the sole effect external to each system.

As in mechanics, work can be a positive or negative quantity. We shall adopt the following convention which is now universal;

work done by a system is positive;
work done on a system is negative.

To amplify on the above definition, let us consider the following situations where work is done in the thermodynamical sense, though not necessarily in the mechanical sense.

(1) Consider a system comprising a container filled with water as used in Joule's classical paddle-wheel experiment. Work is clearly done on the system if we rotate the paddle-wheel by means of descending weights, as performed by Joule, since what happens external to the system is the change in level of a weight.

(2) If we use a motor to rotate the paddle-wheel instead, we may still argue that work is being done on the system since the rotation of the shaft could have been brought about by a falling weight as in (1).

(3) If now we replace the paddle-wheel and spindle altogether by a resistive coil of wire with leads passing through the system boundary and connected to a battery. By such means we are able to bring about the same changes as occurred in (1). Is this work? The answer to this question cannot be provided by the purely mechanical definition. However, with our extended definition, we can show that work is being done, as follows: Let us consider the container with its mass of water and the battery as two interacting systems. If we short-circuit the resistance in the water and connect a motor in between these two systems (note that the motor is external to both), we could repeat the same phenomenon as occurred earlier, viz, the flow of electrons through the lead wires, and have the earlier result, viz, the change in state of the water, reduced to the change in level of a weight as the sole effect external to both systems. (Of course, we assume that the motor is 100 percent efficient! But the side effects lie with an imperfect motor rather than imperfect argument!)

This last illustration leads to the following important result: a flow of electricity *across* the boundary of a system represents a work interaction.

If we were to further modify our above system by having the resistance coil wound around the container on its outside wall, rather than immersed in the water as previously (see Figure 1.1a), by the above argument, work is done if we consider the system comprising the container and the coil as a whole. Now let us consider the system consisting of the mass of water in the container only, as in Figure 1.1(b); the argument of (3) above shows that work cannot be involved since the wires do not lead through the system boundary. What then is it that changes the state of our system? One point we note is that the water changes

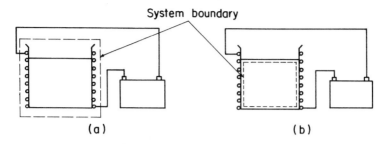

System boundary

(a)   (b)

Figure 1.1

its state because it is 'colder' than the coils while the current is flowing. We could have brought about the same changes in the water by means of an alternative 'hotter' object such as a bunsen burner. (The terms hot or cold as used here are in their usual sense.) If between the two systems, one consists of the water and the other the hot object (either the resistance coil or the burner), we were to interpose a 'heat engine', we would invariably find that the weight-raising cannot be the sole effect external to these systems. There were always other side effects; either the water does not return to its initial state after passing through the engine or the surroundings become warmer in the process. Such a mode of changing the state of a system without work being done has been conventionally referred to as being brought about by *heat*. Though we accept this term in describing the above interaction between our two systems, we are immediately forced to forego the conventional idea that heat is a form of energy which can be added to or withdrawn from the system (the remnants of the caloric theory). Just as work is not a property of a system, neither is heat. From the above considerations, we arrive at the following definition for heat

Heat is an interaction between systems which occurs by virtue of their *temperature* difference when they are brought into contact with each other.

By temperature we mean the thermal state of a system considered with reference to its power of communicating heat to other systems. This latter definition for temperature was coined by Maxwell. Inspite of its qualitative nature and its seemingly circular reference to heat, it is sufficient for our present understanding of the subject. The sign convention for heat is as follows;

heat 'flow into' system is positive;
heat 'flow out of' system is negative.

Before we close our discussion on work and heat, we shall sum up some of their important properties. Heat and work are not forms of energy; both are transient boundary phenomena which exist during the interaction only. Also, heat is not always necessary to cause a temperature rise, as both heat and work have the same end-effects.

Units of work

The unit for work can be inferred through its definition in mechanics. Since in SI units, the unit of force is the newton and the unit of length the metre, the unit of work is the newton-metre. This composite unit has also been called the joule. The same unit is used in thermodynamics.

Units of heat

In SI units, the unit of heat is also the joule. It is not immediately obvious why this should be so. This question will be answered when the formulation of the first law is discussed.

1.4 STATE OF EQUILIBRIUM

The concept of *equilibrium* is central to the study of thermodynamics. By equilibrium, we mean that no further perceptible changes can occur no matter how long we wait. This can be inferred from our definition of property. Since the value of a property changes if the state of the system is altered, property has meaning only when the system is no longer changing. As in mechanics, we distinguish four types of equilibrium. However, for our present purpose, there is no need for us to go into details of these. It suffices for us to say that a system is said to be in a state of equilibrium if after any slight temporary change in the surroundings it returns to its initial state. From this definition, it follows that a finite change from the equilibrium state cannot occur without leaving a permanent change of state in the surroundings. Such states are also known as stable states of equilibrium.

Without further definitions on the internal conditions of a system or assistance from certain fundamental postulates, we are not yet in a position to investigate the conditions of equilibrium within a system. The discussion is therefore confined at present to the equilibrium conditions existing between two systems in equilibrium.

If we take two isolated systems and allow them to come into equilibrium separately, and then bring them into communication with each other, there are two possible interactions between these two systems, viz. work and heat. Consider first that the surface of contact is an *adiabatic wall*, i.e. a surface across which only work interaction is allowed; if the pressures in the two systems on the two sides of this wall is unequal, an unbalanced force exists across this surface. Its magnitude is given by $p_1 A - p_2 A$. The adiabatic wall will move until it reaches a position where this unbalanced force vanishes. Thus, the condition of equilibrium for work interaction across two isolated systems to cease is that the pressure be equal on the two sides of their common adiabatic wall.

If now we replace the adiabatic wall by a *diathermal wall*, i.e. a surface across which only heat interaction is allowed, we find in general heat interaction occurs until the temperature difference across the diathermal wall becomes zero. Thus, the condition of equilibrium for heat interaction across two isolated systems to cease is that the temperature be equal on the two sides of their common diathermal wall.

Note that at this stage we do not forbid the occurrence of temperature or pressure gradients within a system. Under such conditions, the equilibrium conditions discussed above are still valid provided that the same type of gradients occur on both sides of the surface separating the two systems.

1.5 REVERSIBILITY

Consider a system in equilibrium undergoing a certain change of state. We can determine the change in the properties of this system after it is allowed to settle in a new state of equilibrium. From Corollary (3) of our basic definitions, we are able to bring about the same changes by splitting the process up into stages. In order to describe the state of the system after any intermediate stage, we need to know its properties right after that stage. Thus, we require that if a complete description of the property at each point of the process be given, the system must be in a state of equilibrium throughout the process. Such a process is called a *reversible* process. It can only take place infinitely slowly since time has to be allowed for the system to reach equilibrium after an infinitesimally small interaction. As the system is in equilibrium at all times, forces exerted on the system by its surroundings must be balanced by internal forces within the system such that the direction of the process can be reversed and the initial state regained without any other change in the system or surroundings occurring. Because of this, we can also state the following alternative definition for a reversible process:

A process is *reversible* if means can be found to restore the system and its surroundings to their respective initial states.

This definition is very useful when we want to examine whether any given process is reversible or not. The method for such investigations will be given later, after we have studied the second law.

From the above discussion, it is clear that a reversible process is an ideal one and that all real processes are irreversible. Nevertheless, under laboratory conditions near reversible conditions can often be achieved although the types of experiments are not relevant in the present context. Irreversible processes cannot be retraced without leaving a permanent change in the surroundings.

1.6 PRINCIPLE OF EQUIVALENCE

Thermodynamics is a physical science. It is therefore a pursuit of logic just as any other science. There is a somewhat fundamental difference that distinguishes classical thermodynamics from other sciences. All other sciences depend heavily on mathematics as a language for expression. The principles of classical thermodynamics, on the other hand, are essentially verbal and not meaningful in a mathematical sense. Though these laws can be expressed in terms of certain properties that may be mathematically manipulated, the role of mathematics is far less important here than in other sciences. The logical principle of equivalence should therefore be reviewed briefly before embarking on a formal discussion of our subject matter in later chapters.

Let the set of real numbers be denoted by R. If there is an operation $\in$ on the elements in R such that the following three properties are satisfied;

reflexive property: $a \in a$ for all a in R;
symmetric property: If $a \in b$ for some a and b in R, then $b \in a$;
transitive property: If $a \in b$ and $b \in c$ for some a, b and c in R, then $a \in c$;

then there exists an equivalence relation in R. The relation of equality is an example of such a relation. In fact the concept of equivalence relations may be considered as a generalization of the notion of equality. The most important property of an equivalence relation on a set is that it divides the set into mutually disjoint (i.e. having no common elements) subsets called *equivalence classes*. Each equivalence class containing an element x consists precisely of the elements related to x.

Next, we shall consider the concept of equivalence that is applicable to statements. Two statements are said to be (logically) equivalent if they make exactly the same assertion, i.e., if they express the same proposition. Such equivalence requires that the two statements be true under exactly the same conditions. A complete proof of equivalence consists in showing that the two conditions given below are satisfied. These are;

   (i) if B, then A, or
       if not A, then not B;
  (ii) if not B, then not A

These conditions may be represented using a Venn diagram. The first condition requires that B should be completely contained in A, or in mathematical terms B is a subset of A, as shown in Figure 1.2(a). The second, however, requires that A be contained in B or A is the subset of B. Thus, in order that both must hold at the same time, the boundary of A and B must coincide and the equivalence between A and B is complete. In mathematical language, we say that the first

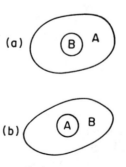

Figure 1.2

condition specifies a sufficient condition, i.e. the occurrence of B is sufficient to guarantee the occurrence of A, and the second a necessary condition, i.e. the occurrence of B is necessary to the occurrence of A. The two conditions taken together give the 'necessary and sufficient conditions' that we so frequently encounter in mathematics. Though this process of showing the equivalence of two statements has the same elegance as that of rigorous mathematical proofs, it will be used only when absolutely necessary so that students not so mathematically (or logically) inclined will not find the text unduly tedious.

EXAMPLES

Example 1.1 Find the work interaction on a gaseous system which obeys the equation of state $pV = mRT$ during (a) an isobaric (constant pressure) process, (b) an isometric (constant volume) process, (c) an adiabatic (zero heat interaction) process and (d) an isothermal (constant temperature) process.
From eqn. (1.2)

$$W_{12} = \int_1^2 F_x \, dx$$

Now, consider the gaseous system to be enclosed in a constant area cylinder closed at one end by a movable piston; as the gas expands the piston is displaced along the axis of the cylinder. The force acting normally on the piston, i.e. in the direction of displacement, is given by

$$F_x = pA$$

where A is the cross-sectional area of the piston or cylinder. On substitution into the above definition

$$W_{12} = \int_1^2 pA \, dx$$

$$= \int_1^2 p \, dV$$

(a) $W_{12} = p \int_1^2 dV = p(V_2 - V) = mR(T_2 - T_1)$

(b) $W_{12} = 0$ since $dV = 0$.

(c) $W_{12} = C \int_1^2 \frac{dV}{V^\gamma}$

$$= C \left[\frac{V^{1-\gamma}}{1-\gamma} \right]_1^2 = \frac{C}{1-\gamma}(V_2^{1-\gamma} - V^{1-\gamma}) = \frac{p_2 V_2 - p_1 V_1}{1-\gamma}$$

$$= \frac{mR(T_2 - T_1)}{1-\gamma}$$

where $pV^\gamma = C$, a constant for an adiabatic process.

(d) $W_{12} = mRT \displaystyle\int_1^2 \frac{dV}{V} = mRT \ln \frac{V_2}{V_1} = mRT \ln \frac{p_1}{p_2}$

Example 1.2 Find the heat interactions for the processes mentioned in Example 1.1 for the same gaseous system.
(a) From elementary definition

$$Q = mc_p (T_2 - T_1)$$

where c_p is the specific heat capacity at constant pressure.
(b) From elementary definition

$$Q = mc_V (T_2 - T_1)$$

where c_V is the specific heat capacity at constant volume.
(c) From definition for adiabatic process

$$Q = 0.$$

(d) Even for this elementary process, it is not possible for us to evaluate the heat interaction based on the definitive approach used for the above three cases.

After the formulation of the first law of thermodynamics, we shall be able to show that for the isothermal process of such a gaseous system

$$Q = W$$

PROBLEMS

1. It is often suggested that thermodynamics is an experimental science; why should this be so?
2. Explain what is meant by the word 'system' in the thermodynamic sense. Describe using a diagram what is meant by the following phrases: a closed system, an open system, a control volume and the control surface.
3. State the sign conventions relating to work and heat interactions between a system and its surroundings.
4. Under what conditions can the expression $W_{12} = \int_1^2 p\,dV$ be equated to a work quantity?
5. Does work depend on path? What about heat?

CHAPTER TWO

Basic Postulates in Classical Thermodynamics and their Consequences

2.1 ZEROTH LAW

If two systems A and B are in equilibrium with system C when in diathermal contact with C, then they are in equilibrium with each other.

Remarks In the statement of the Zeroth Law, it is implicit that the temperature of the systems are uniform throughout. Such hypothetical states are assumed to be reached after a system has been isolated for a sufficient length of time and when there are no internal adiabatic walls to hinder thermal equilibrium within the system. This last condition, however, has not been incorporated into our definition of a system. In practice, for systems consisting of fluids, we usually stir the contents thoroughly before measuring the temperature; this amounts to the measurement of the 'mixing-cup' temperature of the system and represents a spatial average of the non-uniform temperature over the entire system.

The logical content of the Zeroth Law is rather trivial. It follows directly from Maxwell's definition of temperature. If A and C are in thermal equilibrium, then $T_A = T_C$, where T denotes the temperature; similarly, when B and C are in thermal equilibrium, $T_B = T_C$. The result follows from the transitive property of the notion of equality.

Consequences of the zeroth law

The Zeroth Law enables us to classify all systems into a set of equivalence classes such that any two states belonging to one class are equal in temperature, whereas any two states belonging to different classes are unequal in temperature. However, it was not until the enunciation of the second law that the rational measurement of temperature became possible.

2.2 FIRST LAW

When a system executes a cyclic process, the algebraic sum of the work interaction is proportional to the algebraic sum of the heat interaction.

In SI units, the proportionality constant is unity. Thus, we can write this law mathematically as

$$\oint \delta W = \oint \delta Q$$

or

$$\oint (\delta Q - \delta W) = 0 \qquad (2.1)$$

Remarks For students who have not come across it before, the circle placed on the integral sign denotes a cyclic integral, i.e. integrals having the initial and final limits at the same point. Also, the δ notation is used here in place of the differential symbol to indicate that the quantities cannot be evaluated without information on the path taken, i.e. that these quantities are not properties of the system.

Consequences of the first law

By Corollary 6 of our basic definitions, we can define a property E such that

$$E_2 - E_1 = \int_1^2 (\delta Q - \delta W) \qquad (2.2)$$

On writing Q and W for the net quantities of heat and work interactions during the change of state, we have

$$Q - W = E_2 - E_1 = \Delta E \qquad (2.3)$$

This is the mathematical form of the first law that we shall find most useful. The property E is called the energy of the system.

Further remarks

(1) Since E is a state property, the value ΔE can be evaluated using *any convenient path*. The path of the actual process need not be used. The most convenient path is usually a reversible path between the given states because we can easily evaluate Q and W for such a path.
(2) Note that in the formulation of the first law, the energy of the system does not distinguish between heat and work interactions. However, the concept of entropy introduced by the second law later allows us to distinguish between heat and shaft work interactions.
(3) As long as we can integrate the work equation and evaluate Q from fundamental definitions, we do not need the first law. However, this limits us to the consideration of the following four reversible processes only: isobaric, isometric, isothermal and adiabatic processes. If a given process does not follow one of these four paths, the first law is important in the solution of the problem.

Limitations of the first law

The first law deals with the amount of energy (in all its various forms) that are involved in a process. It does not specify the direction of change. There is nothing in the first law to deny the possibility that water will flow uphill or that heat will flow from a region of low to a region of higher temperature, or that gases will unmix.

The criterion for the spontaneity or reversibility of a process is provided by the second law, which specifies the direction of change.

Corollary to the first law The Principle of Conservation of Energy

For an isolated system, $Q = 0$ and $W = 0$, thus $\Delta E = 0$.

Stated in words, we have: The total energy of an isolated system remains constant. This statement of the principle of conservation of energy is a more general statement than the similarly named principle in mechanics. The former takes into account all forms of energy including thermal energy, while the latter considers only mechanical energy in the form of potential and kinetic energy.

Stated in this form, the first law requires that no energy be created or destroyed by an isolated system. This is equivalent to the denial of the idea that a continuously operating device could produce a continuous supply of energy without interacting with its surroundings. Such a device is known as a perpetual motion machine of the first kind, or in short, a PMM1.

2.3 SECOND LAW

Clausius' statement It is impossible for a system working in a cycle to have as its sole effect the transfer of heat from a system at a lower to a system at a higher temperature.

Planck's statement (defines a PMM2) It is impossible to construct an engine which will work in a complete cycle and produce no effect other than raising a weight and exchanging heat with a single reservoir.

Second law and reversibility

It is convenient at this point to re-introduce the concept of reversibility before discussing the various consequences of the second law. This is because the concept of reversibility is closely related to the following development. In Section 1.5 it has been stated that a process is reversible if we can find means to restore the system and its surroundings to their respective initial states. This is, however, easier said than done, as we need a criterion for judging the feasibility of our innovations. This we find in the second law. In the formulations as laid down by Clausius and Planck, the second law clearly indicates the directions of two possible changes and to counter these directions is impossible. Thus, if the hypothetical reversal of a process is found to contradict the second law, the process is said to be irreversible.

It is immediately obvious that heat transfer between systems with a finite temperature difference is irreversible as this contradicts Clausius' statement. The process of heat transfer becomes reversible only if it occurs across an infinitesimally small temperature difference so that by an infinitesimal reduction of the temperature of one of the systems the direction of the interaction is reversed.

In general, the proof for irreversibility will not be as simple. The procedure comprises the devising of a hypothetical cycle making use the reversed process as part of the cycle and invoking the second law to show that this is impossible, i.e. that the device constitutes a PMM2. We shall illustrate this by way of the following example.

Consider the unresisted expansion of a gas. An insulated container is divided into two equal compartments, separated by a membrane. One half of the container is filled with a gas and the other half evacuated. The membrane is suddenly punctured and the gas allowed to fill the entire container. The process is described by

$$p_1 \rightarrow p_2 ; \; p_1 > p_2$$

$$V_1 \rightarrow V_2 ; \; V_1 < V_2$$

and by the first law, the energy of the system remains constant.

Assume now that the process is reversible, and when reversed results in a decrease in volume, an increase in pressure and no change in the energy of the system. We can devise a cycle in which the volume is decreased by a slowly moving piston which does work, and energy equal to this work is returned from a single reservoir via a heat interaction so that the energy of the system is returned to its initial value, and the cycle is complete.

This, however, constitutes a PMM2; hence, the process is irreversible.

One point worthy of note is that the fact that a process is irreversible does not mean that it cannot be reversed. The original state of a system can usually be restored at the conclusion of an irreversible process, though changes would have been incurred in the surroundings.

We shall now return to our main concern with the second law, i.e., the deduction of the various useful concepts and consequences of the second law. To begin with, we shall show the equivalence of the two statements as given by Clausius and Planck. Let us designate these by A and B respectively.

If not A, we can construct an engine $\bar{E}_C$ which transfers heat Q_2 from a cold to a hot *reservoir*. By reservoir, we mean any system in a state of stable equilibrium whose temperature remains constant when it is subjected to finite heat interactions. At the same time we can use an engine E_P operating between the same two reservoirs, to discharge an amount of heat Q_2 to the cold reservoir. Let this engine absorbs Q_1 amount of heat from the hot reservoir. Its net work output per cycle is then given by $W = Q_1 - Q_2$. It is easily realized that the presence of the cold reservoir is superfluous and we can allow the engine E_P to discharge directly into $\bar{E}_C$ instead. Thus, for the system composing of E_P and $\bar{E}_C$ taken together, we have a PMM2. This violates Planck's statement—not B.

On the other hand, if not B, we can construct a PMM2 absorbing Q_1 from a hot reservoir and performing an equal amount of work $W = Q_1$. We can use this work to drive an engine E_C which absorbs heat Q_2 from a cold reservoir and discharges $Q_2 + Q_1$ (by the first law) into the hot reservoir. The net result of the PMM2 and E_C combined, however, is equivalent to an engine transferring heat Q_2 from the cold to the hot reservoir; thus not A.

The proof of equivalence is therefore complete and indeed we can regard the Planck and Clausius statements as corollaries of each other. Further useful corollaries of the second law are given below.

Corollary 1 The efficiencies of a work-producing heat engine operating between two systems at stable states is always less than unity, i.e. $\eta_{he} < 1$.

This corollary is important in heat engine analysis. Also, it introduces several new terms which we shall first define before proceeding to prove its validity.

Heat engine A system which operates in a cycle while only heat and work interactions cross its boundaries.

Stable state A state of stable equilibrium.

Efficiency of heat engine The ratio of work delivered to the heat received from the hot reservoir, i.e.

$$\eta = \frac{W}{Q_1} = \frac{Q_1 - Q_2}{Q_1}$$

Proof Assume that a work-producing heat engine can be devised for which $\eta > 1$. By definition of η, we have

$$\frac{Q_1 - Q_2}{Q_1} > 1$$

i.e.

$$\frac{Q_2}{Q_1} < 0 \qquad (a)$$

Also, since $W > 0$

$$Q_1 - Q_2 > 0 \qquad (b)$$

Inequality (a) requires that Q_1 and Q_2 be of opposite sign, and inequality (b) specifies that $Q_1 > Q_2$, thus

$$Q_1 > 0 \text{ and } Q_2 < 0$$

16

Therefore, heat must flow to the engine from both the hot and cold reservoirs. We may allow Q_2 to flow from the hot reservoir to the cold one so that it suffers no change of state, serving merely as a conductor of heat. The resultant system comprises a PMM2 just as it is for $\eta = 1$, since Q_2 then equals 0.

Since $\eta \geq 1$ violates the Second Law (Planck's statement), the theorem is proved.

Corollary 2 (second law in terms of reversible cycles). For any reversible heat engine which may exchange heat with a single system in a stable state, the net work and net heat in a cycle are zero; i.e.

$$\oint_{\text{rev}} \delta W = \oint_{\text{rev}} \delta Q = 0$$

This result can be shown to be equivalent to Planck's statement using the 'if not A then not B, and if not B then not A' type of proof as demonstrated for the equivalence of the second law statements earlier. The proof is left to interested students.

Consequences of the second law

(i) *The concept of thermodynamic temperature*

This is perhaps the most important consequence of the second law in terms of the rational development of thermodynamics. From now on, the property of temperature, a central concept in thermodynamics, can be quantitatively defined, though only in terms of a hypothetical concept known as an ideal or reversible heat engine. We shall see how this can be translated into practical thermometry later.

This definition of temperature, which is independent of the nature of the thermometric substances, can be brought about in two steps. First, we shall prove the following theorem.

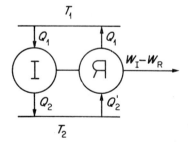

Figure 2.1

Carnot's theorem No engine can be more efficient than a perfectly reversible engine working between the same temperatures, i.e.

$$\eta_{\text{I}} < \eta_{\text{R}}$$

where η_I is the efficiency of the irreversible engine, and η_R of a reversible engine.

Proof Assume $\eta_I \geq \eta_R$

Let W_I be the work done by the irreversible engine I when it receives heat Q_1 and W_R be the work done by the reversible engine R when it receives heat Q_1.

By definition of η

$$W_I \geq W_R$$

Now, reverse the operation of R so that the hot reservoir receives zero net heat and the combined system I–T_1–R exchanges heat with a single reservoir only.

If $W_I - W_R > 0$, this is a PMM2.

If $W_I - W_R = 0$, the cold reservoir suffers no change, and the process is reversible, i.e. I is reversible also.

Since neither alternative is acceptable, Carnot's theorem is proved.

We shall first consider the following corollaries of Carnot's theorem before proceeding to define a thermodynamic temperature scale.

Corollary 1 All reversible engines operating between the same two reservoirs have the same efficiency.

Corollary 2 The efficiency of a reversible heat engine depends solely on the temperatures of the reservoirs between which it is operating.

Both these two corollaries are trivial consequences of Carnot's theorem. The first can be proved by substituting for I a second reversible engine R', while the second follows from the first: Since the efficiency of reversible engines is the same, it should not depend on individual peculiarities of the engine but on some parameters common to all, i.e. the temperatures of the reservoirs.

Now, since the efficiency of a heat engine is defined as

$$\eta = \frac{Q_1 - Q_2}{Q_1} = 1 - \frac{Q_2}{Q_1}$$

it follows from Corollary 2 of Carnot's theorem that Q_2/Q_1 must also be a function of the reservoir temperatures only; thus

$$\frac{Q_1}{Q_2} = f(T_1, T_2)$$

If we assume that one reservoir be at a fixed temperature T_f, say, then the quantity Q/Q_f depends only on the temperature T of the other reservoir whose temperature can be varied, i.e.

$$\frac{Q}{Q_f} = f(T)$$

Alternatively, we may write the inverse functional relation as

$$T = F\left(\frac{Q}{Q_f}\right)$$

It is therefore seen that the thermodynamic temperature can be defined by an arbitrary selection of the form of the function F. For convenience, F should be monotonic, otherwise we would get several values of T for the same value of Q/Q_f, or *vice versa*. The simplest monotonic function, selected by Kelvin, is a direct proportional or linear relation, such that

$$T = T_f \frac{Q}{Q_f} \tag{2.4}$$

The temperature thus defined is known as the thermodynamic temperature. Any other selection for F also defines a thermodynamic temperature. Later on, we shall see that there are certain logical inconsistencies in the Kelvin scale, which the Ramsey scale would bypass. But, for practical purposes, we can make use of only one scale as a standard, and its choice should exclude others to avoid confusion. The general practice now is to adopt the Kelvin scale for temperature measurements due to its ease of accommodating the practical thermometry developed long before the second law was formulated.

Carnot efficiency of a reversible heat engine

As a direct consequence of the Kelvin temperature scale, the efficiency of a reversible heat engine can be written as

$$\eta_{rev} = \frac{T_1 - T_2}{T_1} \tag{2.5}$$

This is known as the Carnot efficiency and is in agreement with Corollary 1 of Carnot's theorem.

The concept of zero thermodynamic temperature

The Kelvin scale of temperature can be rewritten in the following form

$$\frac{T}{Q} = \frac{T_f}{Q_f} \tag{2.6}$$

for which the left and right hand sides are functions of the unspecified and specified reservoirs respectively. They must each be equal to the same constant, the value of which is fixed once we have defined the temperature of the specified reservoir and measured the heat interaction with the same. It therefore follows that $T \to 0$ as $Q \to 0$.

The above argument demonstrates the existence of a zero thermodynamic temperature on the Kelvin scale; however, whether this can be attained in practice is not for the second law to predict. This has to be left to the third law.

The definition of the unit of thermodynamic temperature

With zero thermodynamic temperature hypothetically defined, it is only necessary to further assign an arbitrary value to some fixed temperature level to completely specify the unit of thermodynamic temperature.

The following definition for the unit of thermodynamic temperature was adopted by the Thirteenth General Conference on Weights and Measures at Paris in October 1967:

The unit of thermodynamic temperature is designated by the name 'kelvin' (symbol K). It is the fraction 1/273·16 of the thermodynamic temperature of the triple point of water.

Thus if T_f is taken as the triple point of water in the Kelvin scale, and Q and Q_f are measured as the heat interactions of the engine with the reservoirs, T may be calculated from the following relation

$$T = 273 \cdot 16 \, \frac{Q}{Q_f} \, K \tag{2.7}$$

Practical realization of the thermodynamic temperature

While it is clear from the foregoing discussion that it is conceptually possible to determine the thermodynamic temperature in kelvins at any given temperature level, it remains to be shown how this temperature level can be realized in practical thermometry without the use of the hypothetical reversible heat engine. In what follows, we shall proceed to show that the *ideal gas* temperature is identical to the thermodynamic temperature.

An ideal gas is a gas which obeys the equation of state

$$pv = Rt \tag{2.8}$$

Now let us take this gas as the working substance of a Carnot engine operating between temperatures $t + dt$ and t. A Carnot engine is a heat engine which executes the following four reversible processes in cyclical sequence;

(1) A reversible isothermal process in which the working substance absorbs heat Q_1 from the hot reservoir with the engine performing work.
(2) A reversible adiabatic process in which the temperature of the working substance decreases from the temperature of the hot to that of the cold reservoirs while the engine still doing work.
(3) A reversible isothermal process in which the working substance rejects heat Q_2 to the cold reservoir with work done on the engine.
(4) A reversible adiabatic process in which the working substance is being furthered compressed so that its temperature returns to that of the hot reservoir.

These four reversible processes comprise what is known as a Carnot cycle, shown in Figure 2.2.

20

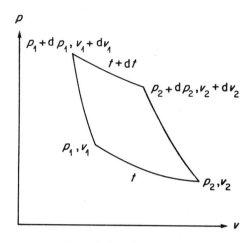

Figure 2.2 A Carnot cycle

The work done on a gas when its state changes from 1 to 2 is given by

$$W = \int_1^2 p\,dv$$

For an isothermal process, this can be integrated with the ideal gas law to give

$$W = p_1 v_1 \ln \frac{v_2}{v_1} \qquad (2.9)$$

In addition, for an isothermal process,

$$W = Q$$

thus

$$Q_1 = R(t + dt) \ln \frac{v_2 + dv_2}{v_1 + dv_1} \qquad (2.10)$$

and

$$Q_2 = -Rt \ln \frac{v_1}{v_2} = Rt \ln \frac{v_2}{v_1} \qquad (2.11)$$

By considering the adiabatic processes, we have

$$\frac{t}{t + dt} = \left(\frac{v_2 + dv_2}{v_2}\right)^{\gamma - 1} = \left(\frac{v_1 + dv_1}{v_1}\right)^{\gamma - 1} \qquad (2.12)$$

Hence

$$\frac{v_2 + dv_2}{v_1 + dv_1} = \frac{v_2}{v_1} \qquad (2.13)$$

and

$$\frac{Q_1}{Q_2} = \frac{t + dt}{t} \qquad (2.14)$$

But, by definition of the Kelvin scale, we have

$$\frac{Q_1}{Q_2} = \frac{T + \mathrm{d}T}{T}$$

Therefore

$$\frac{\mathrm{d}t}{t} = \frac{\mathrm{d}T}{T}$$

On integrating, we obtain

$$\ln t = \ln T + \ln A \tag{2.15}$$

where A is a constant of integration, or

$$t = AT \tag{2.16}$$

Assigning $A = 1$ enables us to use the ideal gas temperature to represent thermodynamic temperature exactly.

Thermometry

The constant volume gas thermometer has been adopted as the standard for temperature measurements by the International Committee on Weights and Measures in 1887. Readings from this instrument are subject to corrections on account of (a) the expansion of the bulb so that the measurements are not strictly made at constant volume, and (b) the fact that the gas in the connecting tube and above the mark in the manometer is not at the temperature of the bulb. These corrections are small in well-designed instruments and may be estimated by assuming that the gas obeys the ideal gas law. The small percentage error thereby introduced into the corrections affects the final result very little indeed. A more serious correction arises due to the nonlinearity exhibited by a real gas over a wide temperature and hence wide pressure range. This has to be taken into account if temperature readings are to be given as ideal gas temperatures. This requires a very precise knowledge of the behaviour of the gas used. We shall discuss a correction procedure using the Joule–Thomson (Kelvin) effect later.

The direct measurement of thermodynamic or ideal gas temperatures is a rather difficult procedure and is now generally only carried out in National Standards Laboratories for precise calibrations. For practical purposes, other thermometers such as the mercury-in-glass thermometer, resistance thermometers, thermocouples, etc., are used instead. In order that these instruments register the ideal gas temperature, they should be calibrated with a constant volume gas thermometer. For accurate measurements, additional corrections have to be applied if the conditions of use are different from those of calibration.

Besides calibration with the gas thermometer, an alternative procedure can be used by calibration against a secondary standard known as the International Practical Temperature Scale, or IPTS for short. This consists of certain reproducible temperature levels, called the fixed points, together with prescribed means of interpolation between these fixed points. These interpolation equations have been so formulated that an IPTS temperature does not differ from the thermodynamic temperature by more than the experimental inaccuracies of the gas thermometer itself. By this means, the temperature registered on any practical thermometer can be easily converted to the ideal gas or thermodynamic temperature.

(ii) *The existence of the property entropy*

We have seen earlier that the first law gives rise to a property known as energy. We shall show now that the second law likewise gives rise to a new property known as entropy.

Hatsopoulos–Keenan Theorem   The quantity $\delta Q/T$ in any reversible process of a system represents the change in the value of a property of the system.

Proof Consider a reversible heat engine receiving heat δQ_{RX} from a reservoir X and rejecting heat into a reversible system A at temperature T, as shown in Figure 2.3. Now the system A–R comprises a PMM2, thus by Corollary 2 of the second law.

$$\oint_{\text{rev}} \delta Q_{RX} = 0$$

From Kelvin's definition of thermodynamic temperature we have

$$\frac{\delta Q}{T} = \frac{\delta Q_{RX}}{T_X}$$

Since T_X is a constant throughout the cycle

$$\oint_{\text{rev}} \delta Q_{RX} = T_X \oint_{\text{rev}} \frac{\delta Q}{T} = 0$$

i.e.

$$\oint_{\text{rev}} \frac{\delta Q}{T} = 0 \qquad (2.17)$$

By Corollary 6 of our basic definitions, this is sufficient for the existence of a property; denoting this by S

$$\mathrm{d}S = \left(\frac{\delta Q}{T} \right) \qquad (2.18)$$

where S is called the entropy of the system.

(iii) *Clausius' Inequality*

This is the analogue of eqn. (2.17) for an irreversible cyclic process. It can be stated as follows: For an internally irreversible cycle the integral of the ratio of the heat δQ received by the system to the temperature T at which heat is received is always less than zero.

Proof Consider the same configuration as shown in Figure 2.3, where A now

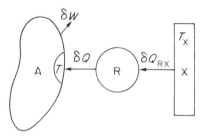

Figure 2.3

represents an irreversible system. When A executes an irreversible cycle. the combined system AR executes an irreversible cycle. The net work of the combined system, according to the first law, is equal to $\oint \delta Q_{RX}$. This net work cannot be positive according to the second law, or AR would be a PMM2 nor can it be zero, for if it was system A and all the elements in its surroundings would return to their original state at the end of the cycle, contrary to the assumption that the process is irreversible. Consequently

$$\oint \delta Q_{RX} < 0$$

But

$$\frac{\delta Q}{T} = \frac{\delta Q_{RX}}{T_X}$$

we therefore have

$$\oint \delta Q_{RX} = T_X \oint_{ir} \frac{\delta Q}{T} < 0$$

Since $T_X > 0$, it follows that

$$\oint_{ir} \frac{\delta Q}{T} < 0 \tag{2.19}$$

(iv) *Entropy and heat in irreversible processes*

Theorem The change in entropy in an irreversible process is greater than the integral $\int \delta Q/T$ over the process, i.e.

$$\int_{ir} dS > \int_{ir} \frac{\delta Q}{T} \tag{2.20}$$

Proof Consider a cycle in which a system begins at state 1, passes through a reversible process R to state 2, and returns by an irreversible process I back to state 1. Since entropy is a property, we have

$$\oint dS = \int_{R1}^{2} dS + \int_{I2}^{1} dS = 0 \tag{2.21}$$

24

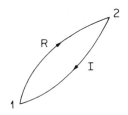

Figure 2.4

By Clausius' inequality

$$\oint_{ir} \frac{\delta Q}{T} < 0$$

therefore

$$\oint \frac{\delta Q}{T} = \int_{R\,1}^{2} \frac{\delta Q}{T} + \int_{I\,2}^{1} \frac{\delta Q}{T} < 0 \qquad (2.22)$$

By definition of entropy

$$\int_{R\,1}^{2} dS = \int_{R\,1}^{2} \frac{\delta Q}{T} \qquad (2.23)$$

Substituting (2.23) into (2.21) gives

$$\int_{R\,1}^{2} \frac{\delta Q}{T} = - \int_{I\,2}^{1} dS \qquad (2.24)$$

Combining (2.22) with (2.24) yields the desired result

$$\int_{I\,2}^{1} dS > \int_{I\,2}^{1} \frac{\delta Q}{T} \qquad (2.20)$$

For small changes in state, we can write this result as

$$dS_{ir} > \left(\frac{\delta Q}{T} \right)_{ir} \qquad (2.25)$$

The above result can be combined with the result for a reversible process to yield the general result

$$dS \geq \frac{\delta Q}{T} \qquad (2.26)$$

where the equality holds for a reversible process.
Corollary (Principle of Increase of Entropy). For an isolated system

$$dS_{isol} \geq 0$$

This follows from the general result, since for an isolated system δQ must be zero.

Isentropic and adiabatic processes

It is obvious from the definition of entropy that the change in entropy in a reversible, adiabatic process in which $\delta Q_{rev} = 0$, is zero. Thus, the term reversible–adibatic implies isentropic, Isentropic, on the other hand, does not imply reversible–adiabatic. For we have

$$dS = 0$$

but since

$$dS \geq \frac{\delta Q}{T}$$

then

$$\frac{\delta Q}{T} \leq 0$$

It follows therefore that for an isentropic process either $\delta Q = 0$ or $\delta Q < 0$, i.e. reversible-isentropic implies adiabatic and irreversible-isentropic implies heat flows out of system whereas for an irreversible-adiabatic process $dS > 0$, i.e. the entropy increases.

This last result is also true for a reversible process in which heat flows into the system; while for a reversible process in which heat flows out of a system, $dS < 0$, i.e. the entropy decreases.

A summary of these various processes is given in the following tables.

Reversible processes		
$\delta Q = 0$	$\delta Q < 0$	$\delta Q > 0$
$dS = 0$	$dS = \frac{\delta Q}{T} < 0$	$dS = \frac{\delta Q}{T} > 0$

Irreversible processes		
$\delta Q = 0$	$\delta Q < 0$	$\delta Q > 0$
$dS > 0$	$dS = 0$	$dS > \frac{\delta Q}{T} > 0$

Second law and unavailable energy

According to the Second Law, when heat is transformed to work by means of a heat engine, the transformation is never complete. A certain amount of heat must be rejected to a heat sink. The lower the temperature of the sink, the smaller will be this unavailable portion. The lowest practical temperature of the sink is that of the surrounding atmosphere or any large body of water, because it does no good to reject heat to any finite body at a lower temperature since the work necessary to maintain a lower temperature would be at least as much if not more than that gained by the use of the lower temperature—this happens only if a reversible engine can be devised.

From the definition of the thermodynamic temperature and the Carnot efficiency, the unavailable energy between two reservoirs is

$$Q_0 = Q_1 \frac{T_0}{T_1} \tag{2.27}$$

where subscript o refers to the sink (cold reservoir) and subscript 1 refers to the source (hot reservoir).

If Q_1 is taken from a finite quantity of working substance of the engine, then a change in temperature of the substance occurs during the heat transfer process and the unavailable energy becomes

$$Q_0 = T_0 \int \frac{\delta Q}{T} = T_0 \, \Delta S \tag{2.28}$$

where ΔS is the entropy change of the working substance between the initial and final states. It is no longer possible to recover this amount of energy as work.

If the surroundings of the system are also taken into account, a similar change in the unavailable energy content occurs in the surroundings. The combined change in the 'universe', used here in the sense of the system together with the parts of the surroundings directly affected by the changes under consideration, is of the form $T_0 (\Delta S + \Delta S_s)$ where ΔS_s is the entropy change in the surroundings. This quantity can be taken as a measure of the irreversibility of the process.

Gibbs equation

An exceedingly important relation bearing the name of Gibbs can be derived by combining the first and second laws of thermodynamics.
From the first law

$$\delta Q = dE + \delta W$$

By the Second Law, for a reversible process

$$\delta Q = T dS$$

Hence, we can write for a reversible process

$$T dS = dE + \delta W_{rev} \tag{2.29}$$

where δW_{rev} is the reversible work interaction during the same process. We shall refer back to this equation later.

2.4 THE STATE PRINCIPLE

The stable state of a system bounded by a fixed surface and subjected to fields prescribed by the surroundings is fully determined by its energy.

Remarks As a system consists of a collection of matter, the properties of a system are essentially the properties of the matter therein contained. Having expressed the laws of thermodynamics in the form of the Gibbs equation, some of the terms occurring in this equation must now be evaluated if the remaining or desired quantity is to be computed for a particular process. In order that this can be done, we must be able to formulate those derived properties occurring in the Gibbs equation as functions of some other independent parameters. The question must therefore be asked: How many independent properties are there for any arbitrary system? The answer to this is provided by the state principle.

According to the state principle, the quantities necessary to describe the boundary of a system, the force fields to which it is subjected and the energy of the system suffice to determine its stable state.

The quantities necessary to describe the boundary of a system can be obtained by examining the forces exerted by a system on its surroundings across an elemental area on the bounding surface. In general, these forces can be split into two components: one normal to the elemental area and the other in the plane of the element. In the case of a fluid, the force per unit area or stress acting normal to the area is known as the pressure and the force along the surface, the surface tension. In the case of a solid, the stress normal to the surface is known as the normal stress and that tangential to the surface the shear stress. In either case both stresses can result in work being done when the system boundary is displaced. In all cases, using the fundamental definition in mechanics, the work done after the system boundary has suffered an infinitesimal displacement can be expressed in the form

$$\delta W = X dx \tag{2.30}$$

where X is known as a generalized force and dx an infinitesimally small generalized displacement of the system boundary.

Similarly, in the presence of external force fields, work will be done on the system if the field is varied. It can be shown that the work done is also given by equation (2.30).

From the above discussion, it is seen that both the quantities necessary to describe the system boundary as well as the presence of external force fields can be reduced to equivalent work modes. If the work is to be performed reversibly, we can then substitute $\Sigma_i X_i dx_i$ for δW in the Gibbs equation to give

$$T dS = dE + \sum_{i=1}^{n} X_i dx_i \tag{2.31}$$

Comparing this expression to its mathematical analogue for a function of several variables, we can then say that the property S, say, is a function of $n + 1$

variables, where n represents the number of reversible work modes that have to be taken into account for the reversible process under consideration.

Consequently, we can state the following corollaries to the state principle.

Corollary 1 The thermodynamic state of a system in stable equilibrium is fixed when its generalized displacements and its energy are specified.

Corollary 2 The number of independent properties necessary for the description of the stable state of a system is equal to the number of reversible work modes plus one.

2.5 THIRD LAW

The entropy of any finite system in a state of thermodynamic equilibrium approaches zero as its temperature approaches zero.

Remarks The above statement is also known as Nernst's heat theorem. It can be represented mathematically as follows.

For any finite system in equilibrium

$$\lim_{T \to 0} S = 0 \tag{2.32}$$

It is also worthy of note that as the temperature tends to zero, the individual components in a mixture separate spontaneously. A mixture is therefore not in a state of equilibrium at very low temperatures.

It is remarkable that Nernst's theorem took almost half a century to establish itself as a law. A number of discrepancies between calorimetric and spectro-scopically measured values of entropy seem to claim exception to the third law. However, on closer examination these exceptions can usually be explained as isotopic or isomeric mixtures of the same constituent which, by our earlier comment on mixtures, are not in a true state of thermodynamic equilibrium.

Corollary It is impossible to reduce the temperature of any system to absolute zero by a finite number of processes.

Proof From the definition of entropy, we can integrate from $T = 0$ to arbitrary T to obtain

$$S = \int_0^T \frac{\delta Q}{T}$$

since the reference entropy at absolute zero is zero by the third law.

Now, consider a system which undergoes a reversible adiabatic process from state A to state B; then

$$\int_0^{T_A} \frac{C_A \, dT}{T} = \int_0^{T_B} \frac{C_B \, dT}{T}$$

where C is the total heat capacity of the system, which is defined as the quantity of heat required to raise the temperature of the system by one kelvin. If $T_B = 0$ K, we must have

$$\int_0^{T_A} \frac{C_A \, dT}{T} = 0$$

Since C_A is always positive, it is impossible to make the above integral converge to zero. Hence $T_B \neq 0$ and the corollary is proved.

Consequences of the third law

The third law permits the determination of the absolute entropy of every substance, including elements and compounds. These entropy values are all properly tied to a common base and can be used conveniently for analysing chemical reactions.

EXAMPLES

Example 2.1 Two identical bodies having same constant thermal capacity c are used as the source and sink of a reversible heat engine. Show that the maximum work obtainable when they are brought to a common temperature T is

$$2c \left(\frac{T_1 + T_2}{2} - T \right)$$

where $T = \sqrt{(T_1 T_2)}$.

Consider an infinitesimal change when the source is cooled from t' to $t' - dt'$ and the sink heated from t to $t + dt$;
heat interaction from source

$$= c \left\{ (t' - dt') - t' \right\} = - c \, dt' = \delta Q_1$$

heat interaction from sink

$$= c \left\{ (t + dt) - t \right\} = c \, dt = \delta Q_2$$

Now by equation (2.4) for a reversible heat engine

$$\frac{\delta Q_1}{\delta Q_2} = \frac{t'}{t}$$

30

Therefore

$$\frac{dt'}{t'} + \frac{dt}{t} = 0$$

On integrating from the initial to the final state, we have

$$\ln \frac{T}{T_1} + \ln \frac{T}{T_2} = 0$$

i.e.

$$T^2 = T_1 T_2$$

whence

$$T = \sqrt{(T_1 T_2)}$$

Work interaction during the infinitesimal change

$$= \delta Q_1 - \delta Q_2$$

$$= - c(dt' + dt)$$

Since the work interaction for a reversible process represents the maximum work interaction, we therefore have for the entire process

$$W_{max} = - c \int_{T_1}^{T} dt' - c \int_{T_2}^{T} dt$$

$$= 2c \left(\frac{T_1 + T_2}{2} - T \right)$$

PROBLEMS

1. What fundamental property is introduced by the zeroth law?
2. Explain the terms adiabatic and diathermal.
3. Using an example, explain the meaning of the word reversible in the context of measuring a work interaction.
4. Name two properties associated with the second law.
5. In a reversible adiabatic process, is the work greater or less than the work of the corresponding irreversible process? Consider both expansion and compression processes.
6. Given two end points, why is the adiabatic work the same for all paths?
7. Name the interpolation instruments used in making measurements of temperature.
8. A toy car of weight 5 N, contains 0·5 kg of water, Consider the following processes:
   (a) the spring in the car is wound up;

(b) the car travels up an incline until its elevation is 1 m, which represents all the energy stored in the spring;

(c) the car starts rolling down the incline with brakes on and stops at the bottom. The heat from the brakes is dissipated in the water.

For parts (a), (b) and (c), what are the changes of entropy of the car, the water and the universe.

9. A string hangs out of the bottom of a black box of unknown contents. When you attach a weight of 5 N to this string, the weight starts falling with a constant velocity. Does this violate the second law? What conclusion can you draw about the state of the box? When you reduce the weight to 2 N, the weight is pulled up with a constant velocity. Does this violate the second law? What is the state of the box?

A friend comes over and tells you that the box contains a white mouse running inside a drum. The mouse is only capable of pulling up a weight of 3 N. Does this fact alter your earlier conclusions? If so, why? If not, why not?

10. Show that if two bodies of thermal capacities c_1 and c_2 at temperatures T_1 and T_2 are brought to the same temperature T by means of a reversible heat engine, then

$$\ln \ T = \frac{c_2 \ln T_2 + c_1 \ln T_1}{c_1 + c_2}$$

Hence show that the total amount of work obtainable from a reversible heat engine working between a source of thermal capacity c_1 and a sink of thermal capacity c_2, the initial temperatures being T_1 and T_2 respectively, is given by

$$c_1 (T_1 - T) \ln \ \frac{T}{T_1}$$

where T is given as above.

11. Demonstrate that the following processes are irreversible;

(a) motion of a solid with friction;

(b) fluid flow with friction;

(c) heat interaction over a finite temperature difference (use Planck's statement);

(d) flow of electric current through a resistance;

(e) mixing of two unlike gases; (to prove irreversibility, it is necessary to postulate a piston made of a semi-permeable membrane which allows only one type of gas molecule to pass through).

CHAPTER THREE

Thermodynamics of One Component Systems

3.1 SIMPLE SYSTEMS

The science of thermodynamics deals with all physical systems in equilibrium or quasi-equilibrium regardless of complexity. However, more often we are concerned with systems subjected to certain idealizations. This generally takes the form of specification of the relevant reversible work mode of the system under consideration. In most cases, we are predominantly occupied with a system involving only one possible work mode, or a *simple system*. However, this term is also used for systems which involve work done by the normal forces at the slowly moving system boundaries alone. Though such systems should more properly be called *simple compressible systems*, we shall nevertheless conform to common usuage of this term also. A formal definition for a simple system is as follows.

A simple system is defined as any system that is not influenced by capillarity, distortion of solid phases, external force fields and internal adiabatic as well as diathermal walls.

From this definition, it can be readily deduced that the reversible work mode for such a system is due to that of the pressure on the moving boundary such that

$$\delta W_{rev} = p \, dV$$

3.2 ENERGY OF A SIMPLE SYSTEM

In the absence of capillarity, distortion of solid phases and external force fields, the energy of the system is not affected by its position in relation to other systems as long as they do not come into direct interaction. The energy of the system is called its *internal energy* and is denoted by the symbol U. Thus, for a simple system at rest

$$E = U$$

When the system is in bulk motion, it is well known from elementary mechanics

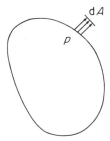

Figure 3.1 A simple compressible system

that the system possesses kinetic energy to the extent of $\frac{1}{2}m\mathbf{V}^2$ where $\mathbf{V}$ is the bulk velocity of the system.

If the system is also subjected to a gravitational field, the effect of this can obviously be taken into account by considering the work interaction on the system when its position is changed within the field. However, due to the fact that the gravitational field is conservative, meaning that the work done is independent of path, it is more suitable to include this into the energy content of the system. It is well known also from elementary mechanics that the potential energy of a system in a gravitational field is given by mgz, where g is the gravitational acceleration and z the height of the centre of mass of the system from a certain reference level. The value of g in SI units is 9.806 m/s^2.

Compounding all these forms of energy together, we can therefore write

$$E = U + \tfrac{1}{2}m\mathbf{V}^2 + mgz \tag{3.1}$$

where E is now known as the *total energy* of the system. For a system of unit mass, its specific total energy is

$$e = u + \tfrac{1}{2}\mathbf{V}^2 + gz \tag{3.2}$$

3.3 GIBBS EQUATION FOR A SIMPLE SYSTEM

On substituting the expression for reversible work into Gibbs equation, we arrive at the following important relation known as Gibbs equation for a simple system

$$T\mathrm{d}S = \mathrm{d}U + p\mathrm{d}V \tag{3.3}$$

Although this equation is derived for a reversible process, all quantities now appearing in it are properties of the system and therefore depend on the end states only. It follows that this expression holds for all processes, reversible or irreversible, connecting equilibrium states of a simple system. However, only in a reversible process will $T\mathrm{d}S$ and $p\mathrm{d}V$ represent the heat and work interactions respectively. In an irreversible process $T\mathrm{d}S > \delta Q$ (this follows from Clausius' inequality) and $pdV > \delta W$ by equal amounts. However, between non-equilibrium states of a simple system

$$dU \neq TdS - pdV \qquad (3.4)$$

It will be seen later that starting with Gibbs equation and expressing the internal energy as a function of V and S, all other thermodynamic properties can be found through differentiation. A relation of this nature is called a *characteristic function*, and embraces all the thermodynamic properties, though in practice it is not possible to begin the formulation of properties of a substance for any one of the characteristic functions because in none of them are all three properties conveniently measurable. Nevertheless, the characteristic function serves a useful purpose as an indicator of the completeness of a formulation. When a formulation has been developed to a point where any one of the characteristic functions can be determined, it is complete.

3.4 OTHER CHARACTERISTIC FUNCTIONS FOR SIMPLE SYSTEMS

We have seen earlier that Gibbs equation defines $U = U(V,S)$ such that U is a characteristic function of a simple system. Instead of U, we can also describe the simple system by means of other characteristic functions, which are functions of other pairs of reference properties, say, S and p, etc. In order that S and p be independent properties of the simple system, we have to replace the differential dV by that of dp. In mathematics, the method of replacing a term ydx by the term $- xdy$ by means of subtracting the differential $d(xy)$ is called a *Legendre transformation*. It is obvious that by this means we can construct two more characteristic functions, namely, by subjecting the other term to Legendre transformation or both terms simultaneously.

Enthalpy

Adding $d(pV)$ to both sides of Gibbs equation, we obtain

$$dU + d(pV) = TdS - pdV + (pdV + Vdp)$$

or

$$d(U + pV) = TdS + Vdp \qquad (3.5)$$

Defining the expression $U + pV$ as the *enthalpy* H of the system, we have

$$H = H(p,S) \qquad (3.6)$$

Equation (3.5) has been referred to as the second Gibbs equation for a simple system.

Helmholtz free energy or work function

If we subject to Legendre transformation the first term of Gibbs equation by subtracting $d(TS)$ from both sides, we obtain

$$d(U - TS) = -SdT - pdV \qquad (3.7)$$

We shall denote by A the function $U - TS$ which is the *Helmholtz free energy* or the *work function*. We see from the above equation that

$$A = A(T,V) \qquad (3.8)$$

Gibbs free energy

If we now subject both the terms on the right hand side of Gibbs equation to Legendre transformations, i.e. by adding $-d(TS) + d(pV)$, we obtain

$$d(U - TS + pV) = Vdp - SdT \qquad (3.9)$$

The function $U + pV - TS$ or $H - TS$ is called the *Gibbs free energy* and denoted by G. From this defining relation, we have

$$G = G(T,p) \qquad (3.10)$$

It is easily seen that all four characteristic functions (u,H,A and G) are properties of the simple system since they themselves are defined by combinations of other properties. The above formulations of characteristic functions are complete because there is no further way to subject the Gibbs equation to Legendre transformation. All these four functions are important thermodynamic properties representing some form of heat or work interaction under particular conditions, as is readily observed from their respective defining relations derived above.

ΔU represents the heat interaction in a reversible isometric (constant volume) process

ΔH represents the heat interaction in a reversible isobaric (constant pressure) process

ΔA represents the work interaction in a reversible isothermal (constant temperature) process

The Gibbs free energy remains constant for a simple system executing a process at constant pressure and temperature. This condition is of particular importance in our consideration of equilibrium of simple systems later.

The property G is also of utmost importance in chemical reactions which usually proceed under constant pressure and temperature. During a chemical reaction, work other than expansion work may also be available from the system. The first law for a chemically reactive simple system now becomes

$$Q = \Delta U + W_e + W_f \qquad (3.11)$$

where W_e is expansion work, and W_f is other useful work, such as electrical work, done by the system
It can be readily shown that

$$\Delta U = Q - W_f \qquad (3.12)$$

at constant volume

$$\Delta H = Q - W_f \qquad (3.13)$$

at constant pressure

$$\Delta A = -W_e - W_f \qquad (3.14)$$

for a reversible isothermal process

$$\Delta G = -W_f \qquad (3.15)$$

for a reversible isothermal process at constant pressure

3.5 MAXIMUM WORK AND MAXIMUM USEFUL WORK

For a system communicating with its surroundings as shown in Figure 3.2, the maximum work that the system can deliver after it comes to equilibrium with the surroundings, can be obtained from the first law as

$$Q - (W + p\Delta V) = \Delta U \qquad (3.16)$$

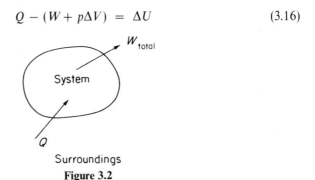

Figure 3.2

where W denotes work done by the system exclusive of expansion work done on surroundings, which are at constant pressure and temperature p and T respectively.

Thus, for a constant volume process

$$W = -\Delta U + Q \tag{3.17}$$

since now $\Delta V = 0$. If the system executes a reversible process, the change of entropy for the universe is zero, whence

$$\Delta S + \Delta S_{sur} = 0$$

Since the temperature of the surroundings remains constant, its change of entropy is given by

$$\Delta S_{sur} = -\frac{Q}{T}$$

Hence

$$\Delta S = -\Delta S_{sur} = \frac{Q}{T} \tag{3.18}$$

Realizing that the work done by the system undergoing a reversible process represents the maximum work that can be delivered by the system, we can therefore write equation (3.17) as

$$W_{max} = -\Delta U + T\Delta S$$

$$= -\Delta(U - TS) = -\Delta A_T \tag{3.19}$$

For a constant pressure process, equation (3.16) becomes

$$W_{max} = -\Delta U - p\Delta V + Q$$

$$= -\Delta(U + pV - TS) = -\Delta G_T \tag{3.20}$$

Stated in words, the maximum work that can be obtained from a system in isothermal contact with the surroundings at all times and with heat interaction occurring at constant volume or pressure, is equal to the decrease in its Helmholtz or Gibbs free energy respectively.

As all systems operating on earth are surrounded by its atmosphere at constant pressure p_o and constant temperature T_o, the maximum work done by such systems is given by

$$W_{max} = -\Delta(U + p_oV + T_oS) \tag{3.21}$$

Since this work equals the total work done by the system less the expansion work done on the surrounding atmosphere, it represents the maximum useful work that the system can perform in contact with the atmosphere. Thus, we can rewrite equation (3.21) as

$$W_{u,max} = -\Delta(U + p_o V - T_o S) \qquad (3.22)$$

where $W_{u,max}$ represents the maximum useful work.

3.6 AVAILABILITY AND IRREVERSIBILITY

The function $U + p_o V - T_o S$ has been defined as the availability function of a system. Its symbol is Φ. Thus, the maximum useful work can be written as

$$W_{u,max} = -\Delta\Phi \qquad (3.23)$$

The actual useful work any system can perform is therefore

$$W_u \leq W_{u,max} \qquad (3.24)$$

The irreversibility of a process is defined as

$$I = W_{u,max} - W_u = -\Delta\Phi - W_u \qquad (3.25)$$

It should be noted that when the system does positive work its availability decreases.

For the system–atmosphere combination, the first law gives

$$Q - W_u = \Delta E_a + \Delta E$$

where subscript a denotes the atmosphere. Now, for the system–atmosphere combination, $Q = 0$; therefore

$$I = -\Delta(E + p_o V - T_o S) + \Delta E_a + \Delta E$$

$$= \Delta E_a + T_o \Delta S - p_o \Delta V$$

But

$$\Delta V_a + \Delta V = 0$$

Therefore

$$I = \Delta E_a + T_o \Delta S + p_o \Delta V_a$$

$$= T_o \Delta(S_a + S) \qquad (3.26)$$

This is the same result as the one derived in Section 2.3.

Example Extend the above treatment to a system exposed to a heat reservoir in addition to an infinite atmosphere. (This problem finds application in power plant analysis, for example.)

For a change of state in a system interacting with the atmosphere and a reservoir X, we have by equations (3.23) and (3.24)

$$W_u \leq - \Delta \Phi - \Delta \Phi_X$$

where Φ_X represents the decrease in availability of the reservoir X. Now

$$- \Delta \Phi_X = - \Delta E_X - p_o \Delta V_X + T_o \Delta S_X$$

But, for the reservoir, the first law gives

$$Q_X = - \Delta E_X - p_o \Delta V_X$$

Since a heat reservoir has a uniform temperature throughout and is capable of absorbing or rejecting any amount of heat without altering its temperature, the changes occurring in it are reversible. Applying Gibbs equation to it, we have

$$Q_X = - T_X \Delta S_X$$

Hence,

$$W_u \leq - \Delta \Phi + Q_X \left(\frac{T_X - T_o}{T_X} \right) \tag{3.27}$$

3.7 RELATIONS BETWEEN PROPERTIES OF SIMPLE SYSTEMS: MAXWELL'S EQUATIONS

Up to this point, we have come across quite a few equilibrium properties that profess to describe a simple system, viz., p, V, T, U, S, H, A, G and the heat capacity. Maxwell has shown that of these properties, the characteristic functions are related to p, V, T and S and that other relations among these properties and other properties also exist. In fact, the number of relations involving the thermodynamic properties and their derivatives are so large that we can only consider the more common and useful ones.

It has been shown in Section 3.3 that Gibbs equation can be written in the form

$$dU = TdS - pdV \tag{3.28}$$

which implies that a functional relationship of the form

$$U = U(S,V) \tag{3.29}$$

exists. Thus, we can write, using the theory of partial differentiation

$$dU = \left(\frac{\partial U}{\partial S}\right)_V dS + \left(\frac{\partial U}{\partial V}\right)_S dV \qquad (3.30)$$

On comparing the two expressions for dU given above, we have

$$\left(\frac{\partial U}{\partial S}\right)_V = T \text{ and } \left(\frac{\partial U}{\partial V}\right)_S = -p \qquad (3.31)$$

If these functions are well behaved (continuous), we have

$$\frac{\partial^2 U}{\partial S \partial V} = \left(\frac{\partial T}{\partial V}\right)_S$$

and

$$\frac{\partial^2 U}{\partial V \partial S} = -\left(\frac{\partial p}{\partial S}\right)_V$$

As the order of differentiation is immaterial, we obtain

$$\left(\frac{\partial T}{\partial V}\right)_S = -\left(\frac{\partial p}{\partial S}\right)_V \qquad (3.32)$$

Similar operations can be performed on other characteristic functions. We shall merely list the main results thus obtained below.

Using the expression for enthalpy, we have

$$dH = Vdp + TdS \qquad (3.33)$$

Hence,

$$\left(\frac{\partial H}{\partial P}\right)_S = V \text{ and } \left(\frac{\partial H}{\partial S}\right)_p = T \qquad (3.34)$$

with

$$\left(\frac{\partial V}{\partial S}\right)_p = \left(\frac{\partial T}{\partial p}\right)_S \qquad (3.35)$$

Starting from the relation for Helmholtz free energy

$$dA = -SdT - pdV \qquad (3.36)$$

yields

$$\left(\frac{\partial A}{\partial T}\right)_V = - S \text{ and } \left(\frac{\partial A}{\partial V}\right)_T = - p \qquad (3.37)$$

together with

$$\left(\frac{\partial S}{\partial V}\right)_T = \left(\frac{\partial p}{\partial T}\right)_V \qquad (3.38)$$

Finally, from the relation for Gibbs free energy

$$dG = - SdT + Vdp \qquad (3.39)$$

we obtain

$$\left(\frac{\partial G}{\partial T}\right)_p = - S \text{ and } \left(\frac{\partial G}{\partial p}\right)_T = V \qquad (3.40)$$

as well as

$$- \left(\frac{\partial S}{\partial p}\right)_T = \left(\frac{\partial V}{\partial T}\right)_p \qquad (3.41)$$

It is therefore seen that all characteristic functions of a simple system are related to the primitive properties p, V and T by the relations

$$\left(\frac{\partial U}{\partial S}\right)_V = \left(\frac{\partial H}{\partial S}\right)_p = T \qquad (3.42)$$

$$\left(\frac{\partial U}{\partial V}\right)_S = \left(\frac{\partial A}{\partial V}\right)_T = - p \qquad (3.43)$$

$$\left(\frac{\partial H}{\partial p}\right)_S = \left(\frac{\partial G}{\partial p}\right)_T = V \qquad (3.44)$$

$$\left(\frac{\partial A}{\partial T}\right)_V = \left(\frac{\partial G}{\partial T}\right)_p = - S \qquad (3.45)$$

The set of equations

$$\left(\frac{\partial T}{\partial V}\right)_S = - \left(\frac{\partial p}{\partial S}\right)_V \qquad (3.32)$$

$$\left(\frac{\partial T}{\partial p}\right)_S = \left(\frac{\partial V}{\partial S}\right)_p \tag{3.35}$$

$$\left(\frac{\partial S}{\partial V}\right)_T = \left(\frac{\partial p}{\partial T}\right)_V \tag{3.38}$$

$$\left(\frac{\partial S}{\partial p}\right)_T = -\left(\frac{\partial V}{\partial T}\right)_p \tag{3.41}$$

is known as Maxwell's relations. By combining these two groups of relations, it is possible to extend or evaluate values of the derived properties when adequate information on the p–V–T relation is known. We shall leave further discussion of this to Chapter 7.

3.8   CRITERION FOR EQUILIBRIUM OF A SIMPLE SYSTEM

We have demonstrated in Section 1.4 that for two systems to be in equilibrium they must have equal temperature and pressure distributions along their common surface. Now, for a simple system, which is not separated by internal adiabatic walls, the same kind of reasoning will show that its temperature must be uniform throughout. Similarly, in the absence of diathermal walls, the pressure at different parts of the simple system must also be equal. If this is not so, work interaction between different parts of the simple system will occur until the pressures of all parts have been equalized.

The above two conditions of uniformity of p and T are necessary but not sufficient for equilibrium. A case in point is a system consisting of two compartments separated by a semi-permeable membrane and having two different gases each at the same pressure and temperature in each compartment.

Since uniformity of p and T are necessary conditions, we shall consider below only those states which have uniform p and T; variations in such states will involve no more than infinitesimal departures from uniformity.

Consider a simple system σ maintained at uniform p and T as shown in Figure 3.3. For any reversible change in an allowed state of σ, the maximum work delivered outside the constant temperature bath B cannot exceed zero if the system σ–B is in a stable state. Since this amount of work is all delivered outside B it does not include work done by σ on the piston. Applying Gibbs equation to the entire system σ–B, we have

$$T\mathrm{d}S - (\delta W_{\mathrm{rev}} + p\mathrm{d}V) = \mathrm{d}U$$

Therefore

$$\delta W_{\mathrm{rev}} = -\mathrm{d}U + T\mathrm{d}S - p\mathrm{d}V$$

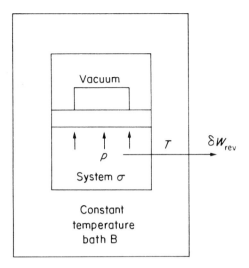

Figure 3.3

Since p and T are constant, this becomes

$$- \, d(U + pV - TS) \leq 0$$

or

$$d(U + pV - TS) \geq 0$$

i.e.

$$dG \geq 0 \tag{3.46}$$

It is therefore necessary and sufficient for equilibrium that all changes to allowed states of this simple system require that

$$dG \geq 0, \text{ at } dp = 0 \text{ and } dT = 0 \tag{3.47}$$

for otherwise the second law would have been violated. It is easy to see that the equality must hold for reversible changes, for if a change from one equilibrium state to another should give

$$dU + pdV - TdS > 0$$

then the reverse would give

$$dU + pdV - TdS < 0$$

and the second state could not have been in equilibrium.

3.9 COEXISTENCE OF PHASES IN A SIMPLE SYSTEM: CLAPEYRON'S EQUATION

The results in the previous section can be applied to the consideration of equilibrium between two phases of the same substance in a simple system. Without loss of generality, let us consider one phase to be liquid, denoted by f and the other gas, denoted by g. The Gibbs free energy for the whole system consisting of masses α of liquid and $(1 - \alpha)$ of vapour is

$$G = \alpha g_f + (1 - \alpha) g_g \tag{3.48}$$

For equilibrium to be maintained under the given p and T

$$dG = 0 \tag{3.49}$$

hence

$$\alpha dg_f + g_f d\alpha + (1 - \alpha) dg_g - g_g d\alpha = 0 \tag{3.50}$$

Since the pressure and temperature for the system has been maintained constant, the specific Gibbs free energies for each *phase* are independent of the amount of substance present in each phase. Therefore, for an allowed variation during which an elemental mass $d\alpha$ passes from one phase to the other

$$dg_f = dg_g = 0$$

Thus

$$g_f = g_g \tag{3.51}$$

i.e. two phases of the same substance can only exist in equilibrium if they possess the same specific Gibbs free energy.

Now consider a slightly different equilibrium state for the two phases at $p + dp$ and $T + dT$; the condition for equilibrium then becomes

$$g_f + dg_f = g_g + dg_g \tag{3.52}$$

Subtracting, we have

$$dg_f = dg_g \tag{3.53}$$

By equation (3.39), this becomes

$$v_f dp - s_f dT = v_g dp - s_g dT \tag{3.54}$$

On rearranging, this becomes

$$\frac{dp}{dT} = \frac{s_g - s_f}{v_g - v_f} = \frac{h_g - h_f}{T_{sat}(v_g - v_f)} = \frac{h_{fg}}{T_{sat} v_{fg}} \tag{3.55}$$

since during a phase change

$$dS = \frac{\delta Q}{T} = \frac{dH}{T_{sat}} \tag{3.56}$$

The quantity h_{fg} is known as the latent heat of vaporization and T_{sat} as the saturation temperature of the substance. This equation, called Clapeyron's equation, links together three commonly measured parameters, viz. the pressure–temperature relation for two phases in equilibrium, the latent heat and the volume increase during a phase change. It is frequently used to compute any one of the three when the other two have been measured, or to compare data obtained from measurements of different kinds.

3.10 REALIZATION OF THE TRUE THERMODYNAMIC TEMPERATURE

In Chapter 2, it has been shown that the thermodynamic temperature can be measured by means of an ideal gas thermometer. In actual practice, the ideal gas approximation cannot be used over the entire temperature range of interest, from close to absolute zero to several thousand kelvins. Real gases exhibit non-linear effects when taken over such a wide temperature range. Nevertheless, it is possible theoretically to correct for the effects due to deviation from the ideal gas behaviour. The procedure for doing this was first conceived and carried out by Kelvin himself, who used for this purpose the equation describing the Joule–Kelvin (Thomson) process, which is

$$\mu_{JK} = \left(\frac{\partial T}{\partial p}\right)_H = \frac{T\left(\frac{\partial V}{\partial T}\right)_p - V}{C_p} \tag{3.57}$$

Assuming that a relation between the gas thermometer temperature t, and the thermodynamic temperature exists, we can introduce into the above equation the quantities measured by means of the gas thermometer as follows. Since

$$\left(\frac{\partial T}{\partial p}\right)_H = \left(\frac{\partial t}{\partial p}\right)_H \frac{dT}{dt} \tag{3.58}$$

$$\left(\frac{\partial V}{\partial T}\right)_p = \left(\frac{\partial V}{\partial t}\right)_p \frac{dt}{dT}$$

and, from the definition of thermal capacities in general

$$C = \frac{\delta Q}{dT} = \frac{\delta Q}{dt}\frac{dt}{dT} = C' \frac{dt}{dT} \tag{3.59}$$

where δQ denotes the total amount of heat absorbed by the system. In particular, we have

$$C_p = C_p' \frac{dt}{dT} \tag{3.60}$$

On substituting into equation (3.57)

$$\left(\frac{\partial t}{\partial p}\right)_H \frac{dT}{dt} = \frac{T\left(\frac{\partial V}{\partial t}\right)_p \frac{dt}{dT} - V}{C_p' \frac{dt}{dT}} \tag{3.61}$$

On rearranging, this becomes

$$\frac{dT}{T} = \frac{\left(\frac{\partial V}{\partial t}\right)_p}{V + C_p' \left(\frac{\partial t}{\partial p}\right)_H} = \frac{\beta_p' \, dt}{1 + \frac{C_p' \mu'_{JK}}{V}} \tag{3.62}$$

where β_p' is the isobaric expansivity of the gas.

The right hand side of the above expression contains only measurable quantities so that the thermodynamic temperature can be found by integration. Obviously, the same gas for which these quantities have been measured should be used for the gas thermometer. The limits of integration extend from the triple point of water T_0, a point defined on the thermodynamic temperature scale, to any arbitrary temperature at which the correction is to be made; thus

$$\ln \frac{T}{T_0} = \int_{T_0}^{t} \frac{\beta_p' \, dt}{1 + \frac{C_p' \mu'_{JK}}{V}} \tag{3.63}$$

The above procedure can be used to obtain thermodynamic temperature from properties measured with any thermometric scale-temperature. For example, if Celsius scale–temperature is used, we can let $t = 0\,°C$ correspond to $T = T_0'\,K$ and then define the unit intervals such that $t = 100\,°C$ corresponds to $T = T_0' + 100$. The integral then becomes

$$\ln \frac{T}{T_0'} = \int_0^t \frac{\beta_p' \, dt}{1 + \frac{C_p' \mu'_{JK}}{V}} \tag{3.64}$$

At $t = 100\,°C$, we have

$$\ln \left(1 + \frac{100}{T_0'}\right) = \int_0^{100} \frac{\beta_p' \, dt}{1 + \frac{C_p' \mu'_{JK}}{V}} = K(100), \text{ say} \tag{3.65}$$

then

$$T'_o = \frac{100}{\exp\{K(100)\} - 1} \qquad (3.66)$$

The number T'_o represents the number of divisions the temperature $T = 0$ K lies below the freezing point of water ($t = 0$ °C). Having thus determined T'_o, equation (3.64) can now be used to compute T for any arbitrary thermometric temperature registered by the thermometer.

The above correction procedure is not restricted to the Joule–Kelvin process alone. Any phenomenon which is both

    1. expressible by a theoretical formula as a function of T, and
    2. measurable in terms of relative scale–temperature t,

can also be used.

Examples

Example 3.1   Calculate the latent heat of steam formed by boiling water under a pressure of 101.325 kN/m². At a pressure near this, a rise of temperature of 1 K causes an increase of vapour pressure of 3.62 kN/m².

Using the Clapeyron equation, equation (3.55)

$$\begin{aligned} h_{fg} &= T_{sat}\, v_{fg} \frac{dp}{dT} \\ &= 373{\cdot}15 \times 1{\cdot}671 \times 3{\cdot}62 \text{ kJ/kg} \\ &= 2257 \text{ kJ/kg} \end{aligned}$$

The specific volume of saturated steam at 101.325 kN/m² is 1.672 m³/kg and that of water 0.001 m³/kg.

Example 3.2   At 273.15 K the specific volumes of water and ice are 0.001000 and 0.001091 m³/kg, and the latent heat of fusion is 334 kJ/kg. Find the melting point increase due to an increase of pressure of 1 atmosphere (101.325 kN/m²).

By the Clapeyron equation, we have

$$\begin{aligned} dT &= \frac{T_{sat}\, v_{sf}\, dp}{h_{sf}} \\ &= \frac{273{\cdot}15 \times (0{\cdot}001000 - 0{\cdot}001091) \times 101{\cdot}325}{334} \text{ K} \\ &= -\,0{\cdot}00753 \text{ K} \end{aligned}$$

Example 3.3   Derive the relation involving the difference of the specific heat capacities of a simple compressible system and discuss the bearing this has on the values of c_p and c_v.

48

Assuming $s = s(v, T)$

$$ds = \left(\frac{\partial s}{\partial T}\right)_v dT + \left(\frac{\partial s}{\partial v}\right)_T dv$$

On multiplying throughout by T, this becomes

$$T ds = T\left(\frac{\partial s}{\partial T}\right)_v dT + T\left(\frac{\partial s}{\partial v}\right)_T dv$$

But

$$c_v = \left(\frac{\partial u}{\partial T}\right)_v = T\left(\frac{\partial s}{\partial T}\right)_v$$

(by 3.42)
whence

$$T ds = c_v dT + T\left(\frac{\partial p}{\partial T}\right)_v dv$$

(by 3.38)

Similarly, by assuming $s = s(p, T)$ and using the appropriate property and Maxwell relations, we have

$$T ds = c_p dT - T\left(\frac{\partial v}{\partial T}\right)_p dp$$

On equating the two expressions thus obtained

$$c_p dT - T\left(\frac{\partial v}{\partial T}\right)_p dp = c_v dT + T\left(\frac{\partial p}{\partial T}\right)_v dv$$

i.e.

$$dT = \frac{T\left(\frac{\partial p}{\partial T}\right)_v}{c_p - c_v} dv + \frac{T\left(\frac{\partial v}{\partial T}\right)_p}{c_p - c_v} dp$$

But

$$dT = \left(\frac{\partial T}{\partial v}\right)_p dv + \left(\frac{\partial T}{\partial p}\right)_v dp$$

whence

$$\left(\frac{\partial T}{\partial v}\right)_p = \frac{T\left(\frac{\partial p}{\partial T}\right)_v}{c_p - c_v}$$

and

$$\left(\frac{\partial T}{\partial p}\right)_v = \frac{T\left(\frac{\partial v}{\partial T}\right)_p}{c_p - c_v}$$

When both these equations are solved for $c_p - c_v$, they yield the same results, namely

$$c_p - c_v = T\left(\frac{\partial v}{\partial T}\right)_p \left(\frac{\partial p}{\partial T}\right)_v$$

Now, by problem 3.3, the partial derivatives are related by the expression

$$\left(\frac{\partial p}{\partial T}\right)_v = -\left(\frac{\partial v}{\partial T}\right)_p \left(\frac{\partial p}{\partial v}\right)_T$$

hence

$$c_p - c_v = -T\left(\frac{\partial v}{\partial T}\right)_p^2 \left(\frac{\partial p}{\partial v}\right)_T$$

From this relation, certain general conclusions can be deduced.

(1) Since $(\partial p/\partial v)_T$ is always negative for all known compressible substances, $c_p - c_v$ can never be negative; hence c_p is always greatr than c_v, except when T or $(\partial v/\partial T)_p$ equals 0.

(2) For a liquid or solid $(\partial v/\partial T)_p$ is usually very small such that the product on the right hand side of the above expression remains relatively small. The difference between the constant pressure and constant volume specific heats of a liquid or solid is, therefore, small. For this reason, many property tables do not distinguish between the specific heat values of liquids and solids. Furthermore $c_p = c_v$ exactly when $(\partial v/\partial T)_p = 0$. This occurs for water at the point of maximum density.

(3) $c_p \to c_v$ as $T \to 0$; at $T = 0$, $c_p = c_v$ exactly.

Example 3.4 Illustration of the use of a characteristic function.

Consider the Helmholtz free energy A; from equations (3.43) and (3.45), we have

$$\left(\frac{\partial A}{\partial V}\right)_T = -p$$

50

and

$$\left(\frac{\partial A}{\partial T}\right)_V = -S$$

Thus pressure and entropy are immediately determined. Since $U = A + TS$ by definition

$$U = A - T\left(\frac{\partial A}{\partial T}\right)_V$$

$$= -T^2\left\{\frac{\partial}{\partial T}\left(\frac{A}{T}\right)\right\}_V$$

If this is differentiated with respect to T

$$C_V = \left(\frac{\partial U}{\partial T}\right)_V = -T\left(\frac{\partial^2 A}{\partial T^2}\right)_V$$

From definition $H = U + pV$, hence

$$H = A - T\left(\frac{\partial A}{\partial T}\right)_V - V\left(\frac{\partial A}{\partial V}\right)_T$$

From definition $G = H - TS$, hence

$$G = A - T\left(\frac{\partial A}{\partial T}\right)_V - V\left(\frac{\partial A}{\partial V}\right)_T + T\left(\frac{\partial A}{\partial T}\right)_V = -V^2\left\{\frac{\partial}{\partial V}\left(\frac{A}{V}\right)\right\}_T$$

To get C_p, the definition $H = U + pV$ is used again to obtain

$$dH = dU + p\,dV + V\,dp$$

Choosing V and T as independent variables, we can write dU as

$$dU = \left(\frac{\partial U}{\partial T}\right)_V dT + \left(\frac{\partial U}{\partial V}\right)_T dV$$

$$= C_V dT + \left(\frac{\partial U}{\partial V}\right)_T dV$$

whence

$$dH = C_V dT + V dp + \left\{ p + \left(\frac{\partial U}{\partial V} \right)_T \right\} dV$$

therefore

$$C_p = \left(\frac{\partial H}{\partial T} \right)_p = C_V + \left\{ p + \left(\frac{\partial U}{\partial V} \right)_T \right\} \left(\frac{\partial V}{\partial T} \right)_p$$

Now since

$$\left(\frac{\partial V}{\partial T} \right)_p = - \frac{\left(\frac{\partial p}{\partial T} \right)_V}{\left(\frac{\partial p}{\partial V} \right)_T}$$

(this can be obtained from relation given in **Problem 3.3**), differentiating the first relation in this example, yields

$$\left(\frac{\partial V}{\partial T} \right)_p = - \frac{\left[\frac{\partial}{\partial T} \left(\frac{\partial A}{\partial V} \right)_T \right]_V}{\left(\frac{\partial^2 A}{\partial V^2} \right)_T}$$

Substitution for this and the other quantities into the above expression for C_p then yields

$$C_p = \frac{T \left\{ \frac{\partial}{\partial T} \left(\frac{\partial A}{\partial V} \right)_T \right\}_V^2}{\left(\frac{\partial^2 A}{\partial V^2} \right)_T} - T \left(\frac{\partial^2 A}{\partial T^2} \right)_V$$

The isothermal compressibility κ_T is defined as

$$\kappa_T = - \frac{1}{V} \left(\frac{\partial V}{\partial p} \right)_T$$

$$= - \frac{1}{V \left(\frac{\partial p}{\partial V} \right)_T} = \frac{1}{V \left(\frac{\partial^2 A}{\partial V^2} \right)_T}$$

The isobaric expansivity β_p is defined as

$$\beta_p = \frac{1}{V}\left(\frac{\partial V}{\partial T}\right)_p$$

$$= -\frac{\left\{\frac{\partial}{\partial T}\left(\frac{\partial A}{\partial V}\right)_T\right\}_V}{V\left(\frac{\partial^2 A}{\partial V^2}\right)_T}$$

The isometric pressure coefficient α_1 is defined as

$$\alpha_V = \frac{1}{p}\left(\frac{\partial p}{\partial T}\right)_T$$

$$= \frac{\left\{\frac{\partial}{\partial T}\left(\frac{\partial A}{\partial V}\right)_T\right\}_V}{\left(\frac{\partial A}{\partial V}\right)_T}$$

We thus see that if the Helmholtz function of a system has been determined over a range of temperatures and volumes then all other thermodynamic properties over that range can be calculated by differentiating the Helmholtz function—this is a property that is true for all other characteristic functions also. We have chosen to illustrate this property for the Helmholtz function because in many problems involving the calculation of properties of a system by the use of statistical mechanics it is most convenient to consider the system as defined by its volume and temperature.

From this example, we see that once a characteristic function has been somehow determined, either through the methods of statistical mechanics or by direct measurements on the system, then thermodynamics permits all other thermodynamic properties to be calculated in a routine manner.

PROBLEMS

1. From the laws of Boyle and Charles, derive the equation of state of a simple compressible system.
2. Given a simple system described by two independent properties x and y, a quantity X is given indirectly by

$$dX = bdx + ady$$

What are the conditions necessary for dX to represent a change in a property?

3. If z is a function of two independent variables x and y, using the expression for the total differential dz show that the three variables x, y and z are related by the expression

$$\left(\frac{\partial x}{\partial y}\right)_z \left(\frac{\partial y}{\partial z}\right)_x \left(\frac{\partial z}{\partial x}\right)_y = -1$$

4. Using the relation given in problem 3 together with the first Maxwell relation

$$\left(\frac{\partial T}{\partial v}\right)_s = -\left(\frac{\partial p}{\partial s}\right)_v$$

deduce the three remaining Maxwell relations.

5. Show that

$$\frac{c_p}{c_v} = \frac{\left(\frac{\partial p}{\partial v}\right)_s}{\left(\frac{\partial p}{\partial v}\right)_T}$$

6. Derive the Clapeyron equation from the Maxwell relation

$$\left(\frac{\partial v}{\partial s}\right)_T = \left(\frac{\partial T}{\partial p}\right)_v$$

7. Derive equation (3.57) for the Joule–Kelvin coefficient.

8. Show that

$$\left(\frac{\partial c_p}{\partial p}\right)_T = -T\left(\frac{\partial^2 v}{\partial T^2}\right)_p$$

$$\left(\frac{\partial c_v}{\partial v}\right)_T = T\left(\frac{\partial^2 p}{\partial T^2}\right)_v$$

9. An insulated cylinder, volume 1 m³ is filled with hydrogen. The cylinder is divided into two equal parts by a conducting porous wall that permits hydrogen to flow slowly from one side to the other. The initial state on one side is given by $p = 1$ atm, $T = 650$ K. On the other side, the gas is at $p = 2$ atm, $T = 530$ K. Demonstrate in a simple manner that this system is *not* in equilibrium on the basis of the temperature difference, on the basis of the pressure difference and on the basis of the difference in Gibbs function. What is the equilibrium state? Demonstrate that, by going from the initial to the final state, the system undergoes an irreversible process.

10. A bottle contains 1 kg of liquid nitrogen at a temperature of 43 K and at atmospheric pressure. What is the least possible cost of this nitrogen? The price of 1 kg of nitrogen at 1 atmosphere and 288 K is $1.00 and the price per kWh of power is $0.05. (Assume $c_p = 1.05$ kJ/kg K and $h_{fg} = 40.8$ kJ/kg.)

11. In a system consisting of 1 kg of steam, initially at 1.5 MN/m² and 533 K,

as it is compressed isothermally to 2.5 MN/m² while being cooled by the atmosphere, find;

(a) the useful work done;
(b) the increase in availability;
(c) the irreversibility of the process.

Assume the temperature of the atmosphere T_o to be 288 K and its pressure p_o to be 0.101 325 MN/m².

12. During the entropy change accompanying a change in the volume and temperature of a simple system, show that

$$T dS = C_V dT + T\left(\frac{\partial p}{\partial T}\right)_V dV$$

Hence, show that the relation

$$S_2 - S_1 = \int_{T_1}^{T_2} C_V d(\ln T) + \int_{V_1}^{V_2} p\alpha_V dV$$

can be used to compute the entropy change between the given end states described by changes in volume and temperature.

13. Derive the following relations which may be used to calculate the temperature changes accompanying a change of state for a simple system executing the various following processes;

(a) constant entropy process

$$\left(\frac{\partial T}{\partial V}\right)_S = -\frac{T}{C_V}\left(\frac{\partial p}{\partial T}\right)_V = -\frac{pT\alpha_V}{C_V}$$

$$\left(\frac{\partial T}{\partial p}\right)_S = \frac{T}{C_p}\left(\frac{\partial V}{\partial T}\right)_p = \frac{VT\beta_p}{C_p}$$

(b) constant enthalpy process

$$\mu_{JK} C_p = T\left(\frac{\partial V}{\partial T}\right)_p - V = V(T\beta_p - 1)$$

where μ_{JK} is the Joule–Kelvin coefficient, defined as $(\partial T/\partial p)_H$

(c) constant energy process

$$\eta_J C_V = p - T\left(\frac{\partial p}{\partial T}\right)_v = p(1 - T\alpha_V)$$

where η_J is the Joule coefficient (for free expansion), defined as $(\partial T/\partial V)_U$.

14. Deduce the following relationships between the two principal heat capacities of a simple system;

$$C_p - C_V = T\left(\frac{\partial S}{\partial V}\right)_T \left(\frac{\partial V}{\partial T}\right)_p$$

$$C_p - C_V = -T\left(\frac{\partial p}{\partial V}\right)_T \left(\frac{\partial V}{\partial T}\right)_p^2 = \frac{VT\beta_r^2}{\kappa_T}$$

$$C_p - C_V = -T\left(\frac{\partial p}{\partial T}\right)_V^2 \left(\frac{\partial V}{\partial p}\right)_T = p^2 V T \alpha_V \kappa_T$$

CHAPTER FOUR

Extension to Flow Process: Thermodynamics of Open System

4.1 FIRST LAW FOR OPEN SYSTEMS

Consider a system Σ and n adjacent masses Δm_i, $i = 1, ..., n$. Let the surface σ of Σ be penetrable to any Δm and consider the process for which these masses are made to merge into Σ, one after another; while Δm_i is crossing σ, the surroundings exert on it a pressure p_i.

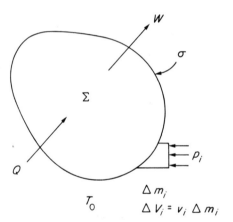

Figure 4.1

Applying the first law to the entire collection of masses before and after the merging process, yields

$$E_2 - (E_1 + \sum_i e_i \Delta m_i) = Q - W + \sum_i p_i v_i \Delta m_i$$

where subscript 2 refers to the final state, after the merging, subscript 1 refers to the initial state, before merging, and $p_i \Delta V_i = p_i v_i \Delta m_i$ represents the compression work done on Σ in merging mass Δm_i into Σ.

On rearranging, this becomes

$$E_2 - E_1 = \sum_i (e_i + p_i v_i)\, \Delta m_i + Q - W$$

In the limit as $\Delta m \to 0$, this becomes

$$dE = \sum_i (e_i + p_i v_i)\, dm_i + \delta Q - \delta W$$

$$= \sum_i h_i^\circ\, dm_i + \delta Q - \delta W \tag{4.1}$$

For simple systems, we have the following property values

$$e = u + \tfrac{1}{2}\mathbf{V}^2 + gz$$

therefore

$$h^\circ = h + \tfrac{1}{2}\mathbf{V}^2 + gz$$

and

$$E = U + \tfrac{1}{2}\mathbf{V}^2 + gZ$$

On dividing equation (4.1) throughout by dt

$$\frac{dE}{dt} = \sum_i h_i^\circ \frac{dm_i}{dt} + \frac{\delta Q}{dt} - \frac{\delta W}{dt} \tag{4.2}$$

Equation (4.2) represents a generalization of the first law to cover systems open to transfer of mass.

The steady flow energy equation (SFEE)

For most engineering systems, a special case of equation (4.2) for which the flow is steady is of special interest. This is because most engineering processes occur under steady conditions. When the flow is steady, the boundary of the system remains unchanged and no expansion work is done by the system on the surroundings. This hypothetical volume which is of fixed size and shape and allows mass flows across its boundaries is called the *control volume*. The work done by the working substance passing through the control volume therefore represents the useful or shaft work W_s done by the system. In addition, when the flow is steady, the accumulation of mass and energy within the control volume is zero. Therefore

$$\dot{Q} - \dot{W}_s = - \sum_i h_i^\circ \dot{m}_i \tag{4.3}$$

where

$$\sum_i \dot{m}_i = 0$$

If we further restrict the number of streams leading into the control volume to two, viz. input and output streams only, the above equation can be written as

$$\dot{Q} - \dot{W}_s = \dot{m}(h^o_{out} - h^o_{in}) \tag{4.4}$$

On dividing throughout by $\dot{m}$, we arrive at one of the most important equations in engineering thermodynamics, the steady flow energy equation (usually written as SFEE for short) i.e.

$$q - w_s = \Delta(h + \tfrac{1}{2}\mathbf{V}^2 + gz) \tag{4.5}$$

where, in keeping with our convention, the heat input per unit mass and the shaft work output per unit mass of the working substance passing through the control volume are written in lower case symbols. Following our nomenclature, these quantities may be called the specific heat input and specific shaft work respectively.

4.2 SECOND LAW FOR OPEN SYSTEMS

Consider the same process as in Section 4.1. The principle of increase of entropy applies to the universe consisting of the system Σ, the n masses and their surroundings which remains at a constant temperature T_o throughout the process. Hence, we have

$$S_2 - (S_1 + \sum_i s_i \Delta m_i) + \Delta S_a \geq 0$$

where ΔS_a denotes the entropy change in the surroundings. On rearranging, this becomes

$$S_2 - S_1 \geq \sum_i s_i \Delta m_i - \Delta S_a$$

In the limit as $\Delta m_i \rightarrow 0$, the above equation can be written as

$$dS \geq \sum_i s_i \, dm_i - dS_a$$

Now, assuming the surroundings act as a reservoir, the heat interaction with the surroundings is reversible, giving

$$dS_a = \frac{-dQ}{T_o}$$

Hence, we obtain on dividing throughout by dt

$$\frac{dS}{dt} \geq \sum_i s_i \frac{dm_i}{dt} + \frac{dQ}{dt} \frac{1}{T_o}$$

i.e.

$$\frac{d(T_oS)}{dt} \geq \sum_i T_o s_i \dot{m}_i + \dot{Q} \tag{4.6}$$

The above result represents a generalization of the second law to systems open to mass transfer. Just like the first and second laws for closed systems, a more useful result can be obtained by combining this with the result of Section 4.1.

Now, from the first law for open systems, we have

$$\frac{dE}{dt} = \sum_i h_i{}^o \dot{m}_i + \dot{Q} - \dot{W} \tag{4.7}$$

Eliminating $\dot{Q}$ between equations (4.6) and (4.7)

$$\frac{d(T_oS)}{dt} \geq \sum_i T_o s_i \dot{m}_i + \frac{dE}{dt} + \dot{W} - \sum_i h_i{}^o \dot{m}_i$$

or

$$\dot{W} \leq -\frac{d(E - T_oS)}{dt} + \sum_i (h_i{}^o \dot{m}_i - T_o s_i \dot{m}_i) \tag{4.8}$$

The quantity $\dot{W}$ represents the total power output from the open system. Part of this is done against the surroundings. Therefore, the useful power or shaft power obtainable from the system is given by

$$\dot{W}_s = (\dot{W} - p_o \frac{dV}{dt}) \leq -\frac{d}{dt}(E + p_oV - T_oS) +$$
$$\sum_i \{(h_i + \tfrac{1}{2}\mathbf{V}_i{}^2 + gz_i) - T_o s_i\} \dot{m}_i$$

i.e.

$$\dot{W}_s \leq -\frac{d\Phi}{dt} + \sum_i (b_i + \tfrac{1}{2}\mathbf{V}_i{}^2 + gz_i) \dot{m}_i \tag{4.9}$$

where

$$b_i = (h_i - T_o s_i)$$

If the flow is steady, the same considerations as stated in Section 4.1 apply and there is no accumulation of availability function within the control volume. Hence

$$\dot{W}_s \leq \sum_i (b_i + \tfrac{1}{2}\mathbf{V}_i{}^2 + gz_i) \dot{m}_i \tag{4.10}$$

Assuming the process to be internally reversible, the equality holds. If we restrict the streams leading into the control volume to input and output streams only, we have

$$\dot{W_s} = \dot{m}\left\{(b + \tfrac{1}{2}\mathbf{V}^2 + gz)_{\text{in}} - (b + \tfrac{1}{2}\mathbf{V}^2 + gz)_{\text{out}}\right\}$$

On dividing through by $\dot{m}$, this becomes

$$w_s = - \Delta(b + \tfrac{1}{2}\mathbf{V}^2 + gz)$$

$$= - \Delta(h + \tfrac{1}{2}\mathbf{V}^2 + gz - T_o s) \tag{4.11}$$

Comparing equations (4.5) and (4.11), it is seen that both give the same specific shaft work when applied to a steady flow isentropic process, as they should. However, when the process is not isentropic, their values naturally differ.

It can be easily shown following the same argument given above that, in the presence of several reservoirs, an additional term of the form

$$\sum_k \dot{Q_k}\left(1 - \frac{T_o}{T_k}\right)$$

should be added to each of the equations after equation (4.7). In particular, equations (4.9) and (4.11) become

$$\dot{W_s} \le - \frac{d\Phi}{dt} + \sum_i (b_i + \tfrac{1}{2}\mathbf{V}_i^2 + gz_i)\dot{m}_i + \sum_k \dot{Q_k}\left(1 - \frac{T_o}{T_k}\right) \tag{4.12}$$

and

$$w_s = - \Delta(h + \tfrac{1}{2}\mathbf{V}^2 + gz - T_o s) + \sum_k q_k\left(1 - \frac{T_o}{T_k}\right) \tag{4.13}$$

Notice that the last term does not change the results of our earlier derivation (equation 4.11) for the system in contact only with the atmosphere, since the setting of $T_k = T_o$ reduces this term to zero.

EXAMPLES

Example 4.1 Application of SFEE: The Joule–Kelvin process

Let us illustrate a simple application of the SFEE by means of the classic experiment of Joule–Kelvin (Thomson) for establishing the dependence of the internal energy of gases on their temperature.

In this experiment, a 'permanent' gas, say air, is being forced steadily through a porous plug constriction in a pipe. The pipe is insulated so that $q = 0$. Also, for the control volume placed across the plug as shown in Figure 4.2 no shaft work is being done, i.e. $w_s = 0$. As the change in kinetic energy of the gas across

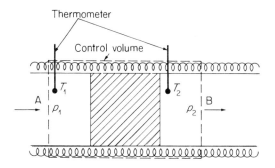

Figure 4.2

the plug and the change in height of the pipe (if any) across the plug are negligible, the SFEE gives

$$\Delta h = 0$$

or

$$h_A = h_B$$

Now if there were no changes in temperature of the gas at A and B, the experiment would have indicated that $h = f(T)$ only. This is because the state of the gas can be determined by only two properties and p and T are the two most conveniently measured properties for this particular experiment. In actual fact, small but definite temperature changes were measured: a decrease of about 0.28 K per atmosphere pressure difference (1 atm $= .101\ 325\ MN/m^2$) for air, and an increase of about 0.028 K per atmosphere pressure difference for hydrogen, both at about 273.16 K initial temperature at A. Thus, the results indicate a small dependence of enthalpy on pressure in addition to the major dependence on temperature. Nevertheless, it remains useful to make the hypothesis that these gases satisfy the relation

$$h = f(T)$$

The theoretical reason behind this will be clear in the next chapter.

Now, since $h = u + pv$, and since pv is known to depend only on temperature, from Boyle's law we thus have

$$pv = f_1(T)$$

and hence

$$u = f_2(T)$$

a function of temperature also.

Example 4.2 Further Applications of SFEE

(a) Adiabatic throttling process

When a fluid flows through a restriction placed in a pipeline, there is an appreciable pressure drop across the restriction in the direction of flow. The flow is said to have been throttled. For such a process, no external shaft work is being performed; if the pipeline is well insulated, no heat interaction occurs either. For such a process, the SFEE becomes

$$0 = \Delta \left(h + \frac{\mathbf{V}^2}{2} \right)$$

Often the flow velocities in a throttled flow are so low that the kinetic energy change term is negligible, hence

$$h_2 = h_1$$

for such a process.

(b) Flow through nozzle or diffuser

A nozzle or a diffuser is a short length of pipe element having varying cross-section such that the flow accelerates or decelerates through the element. Again, no external shaft work is done in the process. Usually the flow through such an element occurs at relatively high speeds, such that there is little time for the fluid to gain or lose energy to the surroundings by means of heat interaction. The process can, therefore, be assumed to be adiabatic. The SFEE again becomes

$$0 = \Delta \left(h + \frac{\mathbf{V}^2}{2} \right)$$

This can be solved for the flow velocity after the fluid has passed through the nozzle or diffuser. The required velocity is given by

$$\mathbf{V}_2 = \sqrt{\mathbf{V}_1^2 + 2(h_1 - h_2)}$$

(c) Heat transfer equipment

Heat transfer equipment normally involves one fluid stream moving relative to another. The two fluid streams are at different temperatures so that the desired result of either cooling one or heating the other can be accomplished in the process. Again, no external shaft work is involved. The flow velocities in such equipments are generally low so that the SFEE becomes

$$\dot{Q} = \dot{m}_c \, \Delta h_c + \dot{m}_h \, \Delta h_h$$

where the dot denotes the rate of the quantity involved and the subscripts c and h refers to the cold and hot streams respectively.

If the heat loss from the hot stream is assumed to be completely carried away by the cold stream, then no heat interaction with the surroundings occurs and $\dot{Q} = 0$. In the case of only one fluid stream flowing through a stagnant fluid mass such as occurring in the cooling panel or evaporator of a refrigerator, the above equation further simplifies to

$$q = h_2 - h_1$$

Students should recognize that these relations are in the forms of the familiar heat balance equations often used in heat transfer courses.

Example 4.3 Charging a high pressure tank. A gas is being pumped into a well insulated tank of volume V. The gas in the tank is initially at pressure p_o and temperature T_o. The gas to be pumped in is at a uniform temperature of T_1 and the inflow rate decreases exponentially with time according to $\dot{m} = \dot{m}_o e^{-at}$. Find the temperature and the pressure of the gas in the tank as a function of time. Neglect the kinetic energy of the incoming gas and assume the gas behaves ideally and has constant specific heats.

Since the tank is well insulated, $\dot{Q} = 0$; there is also no external shaft work, i.e. $\dot{W} = 0$; hence equation (4.2) becomes

$$\frac{dE}{dt} = h^\circ \frac{dm}{dt} = h^\circ \dot{m}_o e^{-at}$$

On integrating and using the condition $E = E_o$ at $t = 0$

$$E = E_o + \frac{h^\circ \dot{m}_o}{a} (1 - e^{-at})$$

where E_o is the initial energy of the tank at the beginning of the charging process. As both the kinetic and potential energies are negligible, we have

$$Mu = M_o u_o + \frac{\dot{m}_o}{a} (1 - e^{-at}) (u_1 + p_1 v_1)$$

Further, from conservation of mass obtained by integrating the mass flow rate equation

$$M = M_o + \frac{\dot{m}_o (1 - e^{-at})}{a}$$

Eliminating M between these two expressions yields

$$\left\{ M_o + \frac{\dot{m}_o}{a} (1 - e^{-at}) \right\} u - M_o u_o = \frac{\dot{m}_o}{a} (1 - e^{-at}) (u_1 + RT_1)$$

hence

$$M_o c_v (T - T_o) = \frac{\dot{m}_o}{a} (1 - e^{-at}) \{c_v (T_1 - T) + RT_1\}$$

i.e.

$$T = \frac{M_o c_v T_o + \dfrac{\dot{m}_o}{a} (1 - e^{-at}) c_p T_1}{\{M_o + \dfrac{\dot{m}_o}{a}(1 - e^{-at})\} c_v}$$

whence

$$p = \frac{MRT}{V} = \frac{R}{V c_v} \{M_o c_v T_o + \frac{\dot{m}_o}{a} (1 - e^{-at}) c_p T_1\}$$

$$= p_o + \frac{\dot{m}_o R}{aV}(1 - e^{-at}) \gamma \, T_1$$

Notice that if the mass inflow is large compared to the initial mass in the tank, the temperature inside the tank tends to γT_1 on fully charged, i.e. as t tends to infinity. If the tank is evacuated initially, the temperature inside the tank becomes independent of time and is equal to γT_1 throughout the entire charging process.

For the case where the charging is effected by a high pressure main, the charging process will stop when the pressure inside the tank reaches that of the main. The charging time can be found by setting $p = p_1$ in the pressure relation; hence the final temperature can be obtained.

Students should note that this example illustrates clearly that classical thermodynamics does not deal with rates of change of a process, since it is quite obvious that the parameter a in the mass flow rate assumption must be separately determined, say, by direct measurement.

If we are only interested in the final state of the process, then the problem can be solved using equation (4.1) instead of equation (4.2). This is left as an exercise for the student (problem 4.3).

PROBLEMS

1. Is the steady flow energy equation true only for frictionless flow or is it true for all types of flow? Explain your answer.
2. If work is obtained at the expense of energy, where does the pv term come from in a flow process?
3. A pressure cylinder of volume V contains air at pressure p_o and temperature T_o. It is to be filled from a compressed air line maintained at a constant pressure p_1 and temperature T_1. Show that the temperature of the air in the cylinder after it has been charged to the pressure of the line is given by

$$T = \cfrac{\gamma T_1}{1 + \cfrac{p_o}{p_1}\left(\gamma \cfrac{T_1}{T_o} - 1\right)}$$

4. Air enters a centrifugal compressor at 101.325 kN/m² and 288 K, and leaves at 200 kN/m² and 370 K. The mass flow rate is 1 kg/s. What power is actually required to drive the compressor? What would the power be for the same pressure ratio if the compression had been reversible and the process assumed to be adiabatic?

5. A diffuser reduces the velocity of an air stream from 270 to 30 m/s. If the inlet pressure and temperature are 101.325 kN/m² and 600 K respectively, determine the outlet pressure. Find also the outlet area required for the diffuser to pass a mass flow of 10 kg/s.

6. A gas flows slowly from a main into an evacuated insulated chamber without doing shaft work during the process. In the final state the chamber contains 1 kg of this gas. What is the change in availability of the gas entering this chamber? Is the change in availability of the chamber the same? If not, what is it? If there is an irreversibility in this process, what causes it? Assume the gas to obey the equation of state $pv = RT$ with $R = 50$ kJ/kg K and $\gamma = 1.4$. The state of the gas in the main is $p = 70$ kN/m² and $T = 500$ K. The atmosphere is at 101 kN/m² and 288 K.

7. One kg of nitrogen at 650 K and 1.5 MN/m² is enclosed in a rigid insulated container. The gas is exhausted to the atmosphere through a reversible adiabatic turbine. How much work does the turbine deliver? Assume nitrogen to obey the ideal gas law and use $c_p = 3.5R$, $c_v = 2.5R$ with $R = 0.297$ kJ/kg K.

CHAPTER FIVE

Properties of Ideal Substances

5.1 INTRODUCTION

So far our discussions have been maintained at a rather general level, mainly outlining thermodynamic principles with a view to their application to closed and open simple systems. In the course, we have defined several thermodynamic properties without any mention of how we can compute their values. In order that the theory developed can be applied to practical problems, means have to be found for evaluating these derived properties as their values are invariably required in estimations of the various parameters of engineering interests. For instance, in the estimation of the reversible heat interaction for a process we need to know the accompanying entropy change. One way of doing this is to idealize the behaviour of certain substances. If our hypothesized models turn out to give a reasonably close description of the actual behaviour of real substances these relations can then be used to calculate the required property values over the range where they apply. Otherwise the properties have to be measured or derived from others measured by means of the thermodynamic relations discussed earlier. Results thus obtained are frequently presented in the form of tables. It is the objective of the present chapter to familiarize students with the handling of property data both in the form of algebraic relations for ideal substances and thermodynamic tables so as to prepare them for engineering cycle analysis and other applications in later chapters.

5.2 IDEAL GASES: SEMIPERFECT AND PERFECT GASES

We start by defining an ideal gas as any homogeneous simple system satisfying the relation

$$pV = n\tilde{R}T \qquad (5.1)$$

where V is the total volume occupied by the gas, n is the number of moles, and $\tilde{R}$ is the universal gas constant, $= 8.3143$ J/K mol. By using the molal volume $\tilde{v}$ of the gas, this may be written as

$$p\tilde{v} = \tilde{R}T \qquad (5.2)$$

Dividing (5.2) through by M, the molecular weight of the gas under consideration, yields

$$pv = RT \tag{5.3}$$

where R is the specific gas constant of the particular gas under consideration.
   Applying the relation

$$\left(\frac{\partial u}{\partial v}\right)_T = T\left(\frac{\partial p}{\partial T}\right)_v - p$$

shows that the internal energy of an ideal gas is a function of temperature only. This can be done as follows

$$\left(\frac{\partial u}{\partial v}\right)_T = T\frac{R}{v} - p = p - p = 0$$

which on integration gives

$$u = u(T) \tag{5.4}$$

From the definition of enthalpy, we have

$$h = u + pv$$

$$= u(T) + RT = h(T) \tag{5.5}$$

Thus, both u and h of an ideal gas are functions of temperature only.
   From the definition for heat capacity, namely $c = \delta q/dT$ in a reversible change under specified conditions, it follows from the definition of the various characteristic functions that an ideal gas has two principal heat capacities defined by the relations (see equation 3.42)

$$c_v = \left(\frac{\partial u}{\partial T}\right)_v \text{ and } c_p = \left(\frac{\partial h}{\partial T}\right)_p \tag{5.6}$$

It therefore follows that for an ideal gas

$$c_v = \frac{du}{dT} = c_v(T) \text{ and } c_p = \frac{dh}{dT} = c_p(T) \tag{5.7}$$

and that

$$c_p - c_v = \frac{d}{dT}(h - u) = \frac{d(RT)}{dT} = R \tag{5.8}$$

Thus, the heat capacities are also functions of temperature only and that their difference equals the gas constant.

Using the heat capacity definitions for an ideal gas, the energy and enthalpy for an ideal gas may then be written as

$$u = \int_{T_o}^{T} c_v \mathrm{d}T \tag{5.9}$$

and

$$h = h(T_o) + \int_{T_o}^{T} c_p \mathrm{d}T$$

$$= RT_o + \int_{T_o}^{T} c_p \mathrm{d}T \tag{5.10}$$

where T_o is the temperature at which u is arbitrarily set equal to zero. Once this is done, equation (5.5) then specifies that the reference enthalpy at T_o must be RT_o.

From Gibbs equation, we have

$$T\mathrm{d}s = p\mathrm{d}v + \mathrm{d}u$$

or

$$\mathrm{d}s = \frac{\mathrm{d}u}{T} + R\frac{\mathrm{d}v}{v} \quad \text{or} \quad \mathrm{d}s = \frac{\mathrm{d}h}{T} - R\frac{\mathrm{d}p}{p}$$

On integration, we obtain expressions for the change in entropy of an ideal gas. These are

$$s - s_o = \int_{T_o}^{T} c_v \frac{\mathrm{d}T}{T} + R \ln \frac{v}{v_o} \tag{5.11}$$

or

$$s - s_o = \int_{T_o}^{T} c_p \frac{\mathrm{d}T}{T} - R \ln \frac{p}{p_o} \tag{5.12}$$

In order that we can proceed to evaluate the integrals in equations (5.11) and (5.12) in closed forms, we introduce further idealization into the behaviour of an ideal gas. This takes the form

$$c_v = \text{constant} \tag{5.13}$$

Any gas satisfying both the ideal gas law, i.e. $pv = RT$ and condition (5.13) at the same time is called a perfect gas. Though it does not seem to be necessary, the name 'semiperfect gas' has been applied to gases which satisfy the ideal gas law and whose c_v is temperature dependent. From now on, we shall continue to make this distinction in the nomenclature of ideal gases and leave the term 'ideal gas' to occasions where heat capacities do not play any explicit part in the discussion.

Thus, for perfect gases, the internal energy, enthalpy and entropy can be computed from the following expressions

$$u = c_v(T - T_o)$$

$$h = c_p(T - T_o) + RT_o$$

$$= c_p T - c_v T_o \qquad (5.14)$$

and

$$s - s_o = c_v \ln \frac{T}{T_o} + R \ln \frac{v}{v_o}$$

$$= c_p \ln \frac{T}{T_o} - R \ln \frac{p}{p_o}$$

$$= c_p \ln \frac{v}{v_o} + c_v \ln \frac{p}{p_o}$$

For changes between two states, 1 and 2 say, these become

$$u_2 - u_1 = c_v(T_2 - T_1)$$

$$h_2 - h_1 = c_p(T_2 - T_1)$$

$$s_2 - s_1 = c_p \ln \frac{T_2}{T_1} - R \ln \frac{p_2}{p_1} \qquad (5.15)$$

$$= c_v \ln \frac{T_2}{T_1} + R \ln \frac{v_2}{v_1}$$

$$= c_p \ln \frac{v_2}{v_1} + c_v \ln \frac{p_2}{p_1}$$

It is indeed remarkable that the perfect gas approximation is valid over wide temperature ranges, thus permitting great simplification in analytic procedures. They are especially useful in gas dynamics where neat closed-form expressions for one-dimensional gas flows can be derived.

Next, we shall turn to the formulation of properties for semiperfect gases. Reference to equations (5.11) and (5.12) shows that this necessitates the evaluation of a definite integral involving one of the principal heat capacities. For this purpose, we define a property ϕ as

$$\phi(T) = \int_{T_o}^{T} c_p \frac{dT}{T} \qquad (5.16)$$

The definition uses c_p rather than c_v because it is more easily measurable. Thus

$$s - s_0 = \phi - R \ln \frac{p}{p_0} \qquad (5.17)$$

and the change between states 1 and 2 is now given by

$$s_2 - s_1 = \phi_2 - \phi_1 - R \ln \frac{p_2}{p_1} \qquad (5.18)$$

For semiperfect gases, values of enthalpy and ϕ have been tabulated as functions of temperature, e.g. in *Gas Tables* by Keenan and Kaye. Also tabulated are the functions relative pressure p_r, and relative volume, v_r. These are defined as

$$\ln p_r = \frac{\phi(T)}{R} \qquad (5.19)$$

and

$$\ln v_r = -\frac{1}{R} \int_{T_0}^{T} c_v \frac{dT}{T} \qquad (5.20)$$

Both these quantities are dimensionless.

These functions are particularly useful for analysis of isentropic processes. Setting $s_2 - s_1 = 0$ in equation (5.18) yields

$$\ln \frac{p_2}{p_1} = \frac{\phi_2 - \phi_1}{R} = \ln p_{r2} - \ln p_{r1} = \ln \frac{p_{r2}}{p_{r1}}$$

i.e.

$$\frac{p_2}{p_1} = \frac{p_{r2}}{p_{r1}} \qquad (5.21)$$

for an isentropic process.

Similarly, it may be shown that

$$\frac{v_2}{v_1} = \frac{v_{r2}}{v_{r1}} \qquad (5.22)$$

for an isentropic process.

5.3 IDEAL LIQUIDS AND SOLIDS

An equation of state can be developed for a liquid or solid by assuming that it is incompressible. Under this assumption, the pressure of the substance can be increased a finite amount by an infinitesimal decrease in volume, which would not result in a significant amount of energy change due to work interaction. Thus, the only means for reversibly changing the internal energy of such a substance is by a heat interaction. It follows therefore from the state principle that only one independent property is sufficient to specify its state (since the reversible work mode is now zero). In the case of a liquid, it should be noted that pressure is still involved from the mechanical point of view and shows up in increased bulk kinetic energy or potential energy due to influence of external force fields. Thus, we may express the internal energy as

$$u = u(T) \tag{5.23}$$

The heat capacity of a liquid or solid is therefore given by

$$c(T) = \frac{du}{dT} \tag{5.24}$$

Integrating between two states, we obtain

$$u_2 - u_1 = \int_1^2 c(T)\,dT \tag{5.25}$$

Since $dV = 0$, Gibbs equation gives

$$dQ = du = T\,ds$$

Hence

$$s_2 - s_1 = \int_1^2 c(T)\,\frac{dT}{T} \tag{5.26}$$

If c is constant, equations (5.25) and (5.26) further reduce to

$$u_2 - u_1 = c(T_2 - T_1) \tag{5.27}$$

and

$$s_2 - s_1 = c \ln \frac{T_2}{T_1} \tag{5.28}$$

For an open system open to transfer of masses, the pressure 'head' is converted into mechanical energies. Under such circumstances, we can define the enthalpy of a liquid as

$$dh = T\,ds + v\,dp = du + v\,dp$$

which gives on integration

$$h_2 - h_1 = c(T_2 - T_1) + v(p_2 - p_1) \tag{5.29}$$

The enthalpy of a liquid (incompressible substance) is therefore seen to be a function of both temperature and pressure, unlike the enthalpy of gases (compressible substances) which depends solely on its temperature.

Equations (5.27) to (5.29) are useful for estimating the properties of subcooled liquids when data of thermodynamic properties are not available. Since the internal energy and entropy are functions of temperature only, the values of u and s for a liquid subcooled to T K will be the same as those of the saturated liquid at T K; and that for h can be computed using equation (5.29) with state 1 on the saturation line at a state determined by either the subcooled temperature or the subcooled pressure of the given state of the liquid. Thus the enthalpy of the subcooled liquid can be obtained from either of the following expressions

$$h_{\mathrm{sub},p} - h_{\mathrm{sat},p} = c\,(T_{\mathrm{sub}} - T_{\mathrm{sat},p}) \tag{5.30}$$

or

$$h_{\mathrm{sub},\,T} - h_{\mathrm{sat},\,T} = v\,(p_{\mathrm{sub}} - p_{\mathrm{sat},T}) \tag{5.31}$$

where the subscripts p and T denote that the saturated state is obtained by holding the pressure or temperature constant respectively.

The reason that two estimated values for h_{sub} can be made, lies in the fact that along the saturated curve pressure and temperature are not independently variable properties; hence, either one can be used to determine a ground state from which the estimation can be made.

5.4 TABLES OF THERMODYNAMIC PROPERTIES

The behaviour of idealized substances in the form of permanent gases, liquids and solids has been considered in previous sections, where it was observed that in general thermodynamic properties have to be presented in the form of tables, although for the special case of perfect gases these can be conveniently represented in the form of algebraic relations. In Section 5.2, some of the parameters commonly tabulated for semiperfect gases were described. In actuality no real substance behaves like a perfect gas under all conditions, although at temperatures well above three times the critical temperature, and at pressures well below 0.2 times the critical pressure all substances begin to show the characteristics of a semiperfect gas. In between these two ranges it is generally necessary to make use of property tables—we shall leave the more involved algebraic forms, as well as other forms of presenting thermodynamic data, to Chapter 7. Our immediate interest in thermodynamic tables is to familiarize students with the use of these tables; their construction and development will be left to Chapter 7 also.

Tables of thermodynamic properties of many substances are available. Among substances of interests to engineers are steam, refrigerants such as ammonia, Freons, etc., common hydrocarbons and liquified gases for cryogenic applications. In general all these tables have the same form. Attention will be concentrated here to the steam tables, primarily because steam is one of the most important engineering substances. Once these are understood, other tables can also be readily used.

Steam tables are divided into two types: saturated water and steam data, and superheated steam data. The manner in which these two types of data are presented is different. They will be considered in turn below.

Saturated water and steam data

Two forms of these data are usually given. One makes use of temperature as the independent variable while the other makes use of pressure. Of course, for saturated states p and T are uniquely related, so that only one form is strictly necessary. However, since both are available, one can choose the most convenient form for interpolation.

It should be pointed out that it is still the practice in steam tables to use the Celsius temperature scale rather than the thermodynamic scale. This should not cause undue difficulty in conversion for the student since the Celsius scale is merely a truncated thermodynamic scale and the two temperature readings are related by the simple expression

$$T = t + 273.15$$

where T is in kelvins and t in °C; the unit interval on both scales have the same magnitude.

Besides pressure and temperature, saturation data tables also contain four other specific properties: v, u, h and s. In tabulating these specific properties, the subscript f is used to denote saturated liquid, g saturated vapour and fg the increase in the property when the state changes from saturated liquid to saturated vapour. Thus, for any property r the following relation holds

$$r_g = r_f + r_{fg}$$

In working with substances close to their saturated states, one frequently encounters mixtures of liquid and vapour. For such two-phase mixtures, the fraction of the total mass that is in the vapour phase is called the *quality* of the mixture, and is usually denoted by x. Since all specific properties are derived from extensive properties, we have by the additive rule

$$R = xMr_g + (1 - x)Mr_f$$

where $R = Mr$. On dividing throughout by M, the total mass of the mixture,

$$r = xr_g + (1 - x)r_f \tag{5.32}$$

$$= r_f + xr_{fg} \tag{5.33}$$

$$= r_g - (1 - x)r_{fg} \tag{5.34}$$

Thus, the quality of a two-phase mixture can be found from any of its known property values by the expression

$$x = \frac{r - r_f}{r_{fg}} \tag{5.35}$$

The quantity $(1 - x)$ has been called the *moisture content* by some writers, though this term is seldom used nowadays.

For values of pressure or temperature intermediate to those tabulated, linear interpolation can be used to obtain the required property values.

Superheated steam data

For superheated steam, both the pressure and temperature are required to specify the state of the steam. As a device for space saving, particularly in abridged tables, the superheat is used instead of the actual temperature of the steam, where superheat means the excess temperature above the saturation temperature corresponding to the pressure of the state.

Specific properties v, u, h and s are listed in more complete versions, though many others in common use do not list u. Due to the dependence of these specific properties on two independent parameters, double interpolation among the four neighbouring states listed in the table is usually required in order that the given state can be determined.

EXAMPLES

Example 5.1 It has been shown in Chapter 2 that unresisted expansion of a gaseous system is an irreversible process. Let us now compute the entropy increase when 1 kg of air at 100 kN/m^2 and 300 K expands into an evacuated insulated container so that its volume is doubled.

By equation (5.15)

$$s_2 - s_1 = c_v \ln \frac{T_2}{T_1} + R \ln \frac{v_2}{v_1}$$

Now since $Q = 0$, and $W_s = 0$, from the first law $\Delta u = 0$ whence $c_v \Delta T = 0$, giving $T_2 = T_1$. Therefore

$$s_2 - s_1 = R \ln 2$$

$$= 0.2871 \ln 2$$

$$= 0.1987 \text{ kJ/kg K}$$

Thus, the entropy increase for 1 kg of air $= 0.1987$ kJ/K

Example 5.2 Estimate the values of u, h and s for Freon–12 at 303.15 K (30°C) and 1.084 MN/m² pressure.

The given state is subcooled as can be seen from Figure 5.1 plotted using the saturation properties given in Thermodynamic Tables: at 303.15 K, p_{sat} = 0.745 MN/m²; and at 1.084 MN/m², T_{sat} = 318.15 K. These two saturated states as well as the given state are shown on the p–h sketch below.

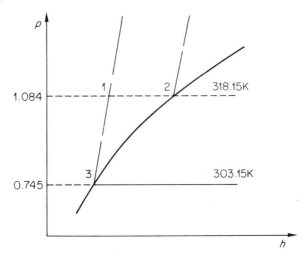

Figure 5.1

Since u and s are functions of temperature only, u and s would have values corresponding to the given temperature, i.e. 303.15 K. From Tables, we find

$$s = 0.2399 \text{ kJ/kg K}$$

Also

$$
\begin{aligned}
u &= h - pv \\
&= 64.59 - 0.745 \times 1000/1304 \\
&= 64.59 - 0.571 \\
&= 64.02 \text{ kJ/kg}
\end{aligned}
$$

To get the value for h, either equation (5.30) or (5.31) can be used. By equation (5.30)

$$
\begin{aligned}
h_1 - h_2 &= c(T_1 - T_2) \\
&= 0.976(303.15 - 318.15) \\
&= -14.65 \text{ kJ/kg}
\end{aligned}
$$

therefore

$$
\begin{aligned}
h_1 &= 79.71 - 14.65 \\
&= 65.06 \text{ kJ/kg}
\end{aligned}
$$

By equation (5.31)

$$h_1 - h_3 = v(p_1 - p_3)$$
$$= \frac{1}{1304}(1.084 - 0.745) \times 10^3$$
$$= 0.26 \text{ kJ/kg}$$

therefore

$$h_1 = 64.59 + 0.26$$
$$= 64.85 \text{ kJ/kg}$$

The discrepancy between these two calculated values for h stems from the fact that v and c values for liquid Freon–12 have not been tabulated in the same detail as the property values for the gaseous state found in abridged tables commonly available for student use.

Example 5.3 A pressure vessel having a volume of 1 m^3 contains 3 kg of water and water vapour mixture at a pressure of 500 kN/m^2. Find (a) the temperature of the contents, (b) the volume and mass of water and (c) the volume and mass of vapour.

(a) Since water and water vapour exist in equilibrium, the mixture is saturated and the temperature therefore equals T_{sat} at $p = 500 \text{ kN/m}^2$, i.e. 424.95 K (or 151.8°C)

(b) Specific volume of mixture $= 1/3 = 0.3333 \text{ m}^3/\text{kg}$
Now since

$$v = v_f + x v_{fg}$$

we have

$$0.3333 = 0.0011 + 0.3737x$$

therefore

$$x = \frac{0.3322}{0.3737} = 0.889$$

The mass of liquid $= 3(1 - x)$

$$= 0.333 \text{ kg}$$

The volume of liquid $= m_f v_f = 0.333 \times 0.0011$

$$= 0.000366 \text{ m}^3$$

$$= 0.366 \, l$$

(c) The mass of vapour $= 0.889 \times 3$

$$= 2.667 \text{ kg}$$

The volume of vapour $= m_g v_g$

$$= 2.667 \, (0.3737 + 0.0011)$$

$$= 2.667 \times 0.3748$$

$$\simeq 0.9995 \text{ m}^3$$

Example 5.4 Discuss the qualitative features of a *T–S* diagram.

Since by definition $dS = \delta Q/T$ for a reversible process, it follows that along an isobar

$$dS = \frac{C_p \, dT}{T}$$

Therefore

$$\left(\frac{\partial T}{\partial S}\right)_p = \frac{T}{C_p}$$

As both T and C_p are positive quantities, the slope of an isobar on the *T–S* diagram is necessarily positive. This slope equation does not apply to the liquid–vapour two phase region because the sensible heat equation

$$\delta Q = C_p \, dT$$

does not apply to an isothermal–isobaric phase change. Instead the entropy change is given by

$$\Delta S = \frac{\Delta H_{fg}}{T}$$

where ΔH_{fg} is the latent heat of the phase change and T the constant temperature at which it occurs. Since T is constant, this part of the isobar on the *T–S* diagram is horizontal. As the value of ΔH_{fg} decreases as the saturation temperature increases, and is zero at the critical point, the two phase region forms a dome-shaped figure which narrows towards the top at a temperature of T_c, the critical temperature. At this point, the value of C_p for the two phases becomes infinite. Thus, the critical pressure isobar has zero slope at the critical point.

In the compressed liquid region, due to the fact that entropy change is essentially a function of temperature only (see equation 5.28), the isobars are crowded onto the saturated liquid line.

We can also deduce the shape of an isenthalpic (constant enthalpy line) by the following reasoning: in the ideal gas (high temperature or low pressure) region, constant temperature implies constant enthalpy. Hence, the isenthalpic is horizontal. In the two phase region, there is a continuous increase in enthalpy as the fluid is being evaporated from the saturation liquid end towards the saturation vapour end. Furthermore, the saturated vapour has higher enthalpy at higher temperature. Thus, the shape of the isenthalpic in the two phase region is concave upwards. By virtue of the continuous nature of the thermodynamic functions, this property of the isenthalpic is expected to extend into the region near the two phase dome.

These features of the T–S diagram are shown in Figure 5.2.

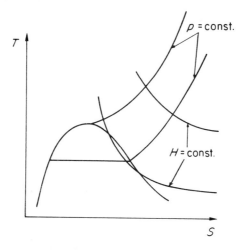

Figure 5.2

Example 5.5 Discuss the qualitative features of the p–H diagram.
Consider the relation

$$\left(\frac{\partial p}{\partial H}\right)_S = \frac{1}{V}$$

The left hand side represents the slope of an isentropic on the p–H diagram. The relation indicates that this equals the density of the substance. As the density is a positive quantity, the slope of an isentropic is necessarily positive. Furthermore, the density of a gas decreases with temperature at constant pressure, and since $H \propto T$, the slopes of two neighbouring isentropics are such that the one at higher temperature has a smaller slope and the isentropics fan out as indicated in Figure 5.3. Unlike the slope equation for an isobar on the T–S diagram, the slope equation for an isentropic on the p–H diagram remains valid in the two phase region. Since density is a smooth thermodynamic property at the two phase boundary, the slope of an isentropic is continuous through the saturation line.

The width of the two phase region is determined by the quantity ΔH_{fg}, which decreases to zero at the critical point. Thus, the two phase region is again a dome-shaped figure narrowing towards zero at the critical pressure.

In addition, we can also deduce the shape of an isotherm on the p–H diagram as follows: In the low pressure region, the gas behaves essentially as an ideal gas for which H is a function of T only. Thus, in this region constant H means constant T, hence the isotherm is perpendicular to the enthalpy axis, or vertical in shape. In the two phase region, the saturation pressure is a function of saturation temperature only; hence constant p means constant T in the two phase region and the isotherms are horizontal. In the compressed liquid region, the two principal specific heats are almost equal with $H \approx U$. Since $H = U + pV$ by definition, we conclude that H is almost independent of p and is essentially a function of T only. Thus, the isotherm is again vertical.

These features of the p–H diagram are shown in Figure 5.3.

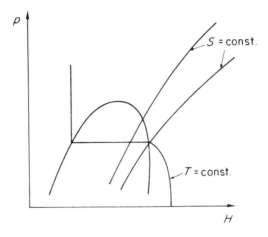

Figure 5.3

Example 5.6 Discuss the qualitative features of the H–S or Mollier diagram.

The Mollier diagram is essentially a distorted version of the T–S diagram— distorted in such a way that the constant enthalpy lines become straight and horizontal. The constant pressure lines in the liquid region merge into the saturated liquid line as before. In the two phase region, the isobars and isotherms coincide. Consider the relation

$$\left(\frac{\partial H}{\partial S}\right)_p = T$$

where the left hand side represents the slope of an isobar on the H–S diagram. From this relation, it is seen that in the two phase region for which the tempera- ture is constant along an isobar, the slope of the isobar and hence the isotherm is constant. In other words, isobars and isotherms are straight lines in the two phase region. In the superheat region, the slope of an isobar remains continuous due to the above relation. However, the isotherms suffers a sharp bend towards

80

the right and their slope decreases asymptotically to zero so that they become perpendicular to the enthalpy axis. This is necessarily so because towards the ideal gas region, constant enthalpy implies constant temperature.

Due to the distortion of the two phase dome, the critical point is no longer at the peak of the dome.

These features of the H–S diagram are shown in Figure 5.4.

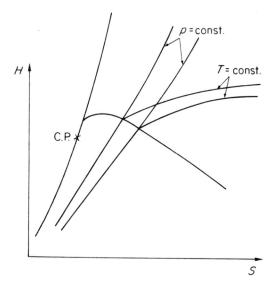

Figure 5.4

PROBLEMS

1. Establish the law governing a reversible adiabatic (isentropic) process for an ideal gas.
2. One kg of steam is enclosed in a cylinder by a piston at 200 kN/m² and 523 K. The steam is expanded adiabatically to 75 kN/m². The piston delivers 10 kJ of work. Is this a reversible or irreversible process? Explain.
3. One kg of steam initially saturated at 5 MN/m² expands reversibly and isothermally to 1 MN/m² in a cylinder. Find the heat and work interactions with the surroundings. If the fluid had been a perfect gas, the heat and work interactions would have been equal because the internal energy being a function of temperature would have been constant. Why is this not so for steam? Why would these interactions be more nearly equal if the steam had expanded from, say, 0.5 MN/m² to 0.1 MN/m²?
4. Steam at 0.3 MN/m² and 423.15 K is expanded through a nozzle to atmospheric pressure of 0.1 MN/m². The steam enters the nozzle with negligible kinetic energy and the flow can be assumed to be reversible. Determine the outlet velocity from the nozzle.
5. One kg of steam at 673.15 K and 1.5 MN/m² is enclosed in a rigid insulated

container. The steam is exhausted to the atmosphere through a reversible adiabatic turbine. Find the work delivered by the turbine. Also determine the maximum work and the maximum useful work derivable from the steam between the two given states. What is the irreversibility of this process? If this is non-zero, how do you account for this?

6. Steam flows slowly from a steam main where the pressure is 3.5 MN/m^2 and the temperature is 650 K into an evacuated bottle, without heat interaction occurring during the process. When the flow has stopped, 1 kg of steam has entered the bottle. Determine;

   (a) the final temperature in the bottle;
   (b) the change in availability of the steam between the state in the main and the final state in the bottle;
   (c) the maximum useful work derivable from the steam in its final state;
   (d) the maximum useful work stored in the steam due to its change of state from before to after the process;
   (e) the maximum shaft work that the steam main must performed onto the steam during the charging process;
   (f) the change in availability of all the steam left behind in the main, assuming the main to be a reservoir of infinite volume;
   (g) the irreversibility in the process.

7. Air flows steadily at a rate of 1 kg/s through a porous plug with negligible inlet and outlet velocities. The inlet condition is 0.1 MN/m^2 and 823 K, the outlet condition is 0.05 MN/m^2 and 723 K. Find (a) the rate of heat interaction, (b) the change in entropy per kg of air passing through the plug using the following assumptions;

   (i) that the air behaves as a perfect gas;
   (ii) that the air behaves as a semiperfect gas, using the mean specific heats for the relevant temperature range in your calculations;
   (iii) that the air behaves as a semiperfect gas, using property values tabulated in *Gas Tables* for your calculations.

CHAPTER SIX

Direct and Reversed Heat Engines

6.1 REVERSIBLE AND IRREVERSIBLE STATE CHANGES ON p–v AND T–s DIAGRAMS

As a prelude to cycle analysis, it is appropriate to first examine the various processes of engineering importance. Since a simple system has two independent properties, a state of a simple system can be represented on a two-dimensional graph with the relevant properties as axes. Such graphs are commonly referred to as a diagram in thermodynamics literature. As the number of thermodynamic properties is not small, a vast combination of such diagrams is possible. Of these, two have been more frequently referred to than others. These are graphs having p–v and T–s axes. It is not difficult for us to realize their relative importance if we recall that $p\mathrm{d}v$ and $T\mathrm{d}s$ represent the reversible work and heat interactions respectively. These quantities are thus visualized as the area under the curve representing the process on the p–v and T–s diagrams.

In this section, we shall summarize the results for the more important processes, together with their paths on the p–v and T–s diagrams. For all cases, we shall illustrate the process for the case of a vapour. i.e. near to the two-phase region. For processes executed by permanent gases, the actual paths consist of only the portions outside the two-phase dome on these diagrams.

As results for both non-flow and flow processes will be given, we shall first derive the general expressions for work and heat interactions for a fluid undergoing a steady flow process. From these general equations, students should have no difficulty in deducing the particular results listed below. The potential and kinetic energies will be neglected in the following derivation.

From the SFEE

$$\mathrm{d}q_{\mathrm{rev}} - \mathrm{d}w_{\mathrm{s}} = \mathrm{d}h \qquad (6.1)$$

By the second Gibbs equation

$$\mathrm{d}q_{\mathrm{rev}} = \mathrm{d}h - v\mathrm{d}p \qquad (6.2)$$

hence, by subtracting (6.1) from (6.2)

$$\mathrm{d}w_{\mathrm{s}} = -v\mathrm{d}p$$

On integration, this gives

$$w_s = - \int_1^2 v\,dp \qquad (6.3)$$

Hence,

$$q + \int_1^2 v\,dp = \Delta h \qquad (6.4)$$

(i)   *Heating at constant pressure (Figure 6.1)*

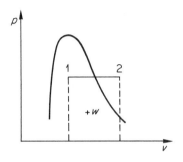

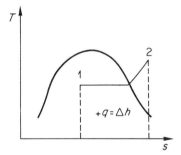

Figure 6.1

Non-flow process

$$w = \int_1^2 p\,dv = p(v_2 - v_1)$$
$$q = \Delta u + \int_1^2 p\,dv = u_2 - u_1 + p(v_2 - v_1) = h_2 - h_1$$

Steady-flow process

$$w_s = 0$$

$$q = h_2 - h_1$$

84

84

(ii) Heating at constant volume (Figure 6.2)

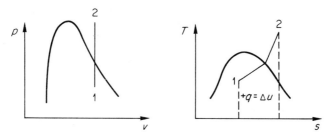

Figure 6.2

Non-flow process

$$w = 0$$

$$q = \Delta u$$

This is not possible as a steady-flow process.

(iii) Isothermal process (Figure 6.3)

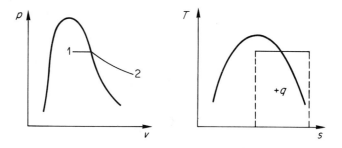

Figure 6.3

Non-flow process

$$q = T(s_2 - s_1)$$
$$w = q - \Delta u$$
$$= T(s_2 - s_1) + (u_1 - u_2)$$

Steady-flow process

$$w_s = - \int_1^2 v\,\mathrm{d}p$$

$$q = \Delta h + w_s = \Delta h - \int_1^2 v\,\mathrm{d}p$$

Notice for the particular case of an ideal gas, heat input = work output and both equal $- RT \ln p_2/p_1$

(iv)   *Isentropic process (Figure 6.4)*

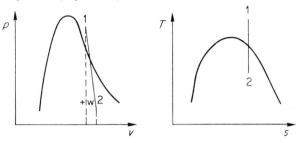

Figure 6.4

Non-flow process

$$q = 0$$
$$w = - \Delta u = u_1 - u_2$$

Steady-flow process

$$q = 0$$
$$w_s = - \Delta h = h_1 - h_2$$

(v)   *Reversible polytropic process: $pv^n = constant$ (Figure 6.5)*

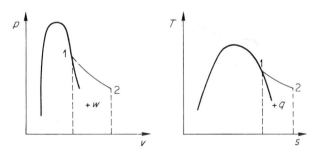

Figure 6.5

Non-flow process

$$w = \int_1^2 p\,dv = \frac{p_2 v_2 - p_1 v_1}{1 - n}$$

$$q = \Delta u + w$$
$$= u_2 - u_1 + \frac{p_2 v_2 - p_1 v_1}{1 - n}$$

Steady-flow process

$$w_s = -\int_1^2 v\mathrm{d}p = \frac{n(p_2v_2 - p_1v_1)}{1 - n}$$

$$q = \Delta h - \int_1^2 v\mathrm{d}p$$

$$= h_2 - h_1 + \frac{n(p_2v_2 - p_1v_1)}{1 - n}$$

(vi)   Throttling process: irreversible adiabatic steady-flow process in which no shaft work is done (Figure 6.6)

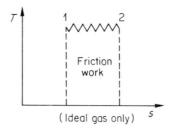

Figure 6.6

From SFEE

$$h = 0 \text{ or } h_1 = h_2$$

When the fluid is an ideal gas, this reduces to

$$T_1 = T_2$$

(vii)   Other irreversible adiabatic process

As a consequence of the second law, the entropy increases during an irreversible adiabatic process. Therefore, all such processes can be shown on the T–s diagram with the final state lying to the right of the initial one. Two such processes are shown in Figure 6.7, one for an expansion and the other for a compression. It is rather obvious that more work is required for compression and less work obtainable from expansion in the presence of irreversibilities, when compared with the insentropic process over the same pressure range. In this respect, the throttling process can be regarded as an expansion which gives maximum increase in entropy. Since these processes are irreversible, no interpretation can be attached to the area under the 'paths'.

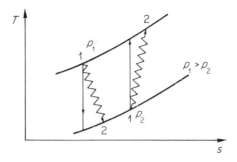

Figure 6.7

Since the laws of thermodynamics do not enable quantitative predictions to be made for irreversible processes, empirical means have to be resorted to, one of which is to define what have come to be known as process efficiencies. We introduce at this stage two such efficiencies dealing with turbines and rotary compressors, since they are rather exclusively used in analysing heat engines. These are known as turbine and compressor isentropic efficiencies. They are respectively defined as

$$\eta_T = \frac{w_{act}}{w_{isen}} \quad \text{and} \quad \eta_C = \frac{w_{isen}}{w_{act}} \tag{6.5}$$

When the working substance can be considered as a perfect gas, these become

$$\eta_T = \frac{T_1 - T_2}{T_1 - T_2'} \quad \text{and} \quad \eta_c = \frac{T_2' - T_1}{T_2 - T_1} \tag{6.6}$$

where the primed state refers to the corresponding state if the process were carried out isentropically.

Note that the efficiencies are always defined such that their values are always less than one. In fact, for all cases it can be easily observed that they conform to the following structure

$$\text{any efficiency} = \frac{\text{what we want}}{\text{what we have to pay}}$$

The process irreversibilities referred to above are also known as internal irreversibilities. They are caused by internal turbulence created by the rapid expansion of the fluid.

There are many other forms of irreversibility. An obvious one is external irreversibility. This also has great influence on the performance of heat engines. It has also been referred to as thermal irreversibility as it is caused by the transfer of heat across a dividing wall as well as the absorption or rejection of heat over a temperature range, rather than from or to constant temperature reservoirs.

88

6.2 VAPOUR POWER CYCLES

In vapour power plants, the working fluid is usually recirculated so as to operate in a closed cycle or as a heat engine. An exception is steam locomotive engines where the expanded vapour is exhausted to the atmosphere so that the engine size can be kept down. These cycles have two characteristics in common.

(a) The working fluid is a condensable vapour which is in the liquid phase during part of the cycle.
(b) The cycle consists of a succession of steady-flow processes, with each process carried out in a separate component specially designed for the purpose. These components are connected in series so that the fluid passes through a cycle of operations after passing through each and every component in turn. The sequence of essential components are (i) boiler, (ii) engine or turbine, (iii) condenser and (iv) feedwater pump. The furnace, though essential to the operation of the plant, is not incorporated into the cycle.

To simplify the analysis of power cycles, the kinetic and potential energies of the fluid between entry and exit of each component is ignored. These are usually very small compared to the enthalpy change.

Practical considerations

(i) *The choice of the plant*

This depends on the following factors;

(a) operating cost: this is a function of the overall efficiency and the types of fuel used;
(b) capital cost: this is a function of size, complexity and type of fuel used.

The overall efficiency is defined as

$$\eta_{overall} = \frac{\text{useful work/cycle}}{\text{heating value of fuel supplied/cycle}} \qquad (6.7)$$

$$= \eta_{comb}\,\eta_{a,c,}$$

where η_{comb} denotes the combustion efficiency and $\eta_{a,c,}$ denotes the actual cycle efficiency.

$\eta_{comb} < 1$: Combustion is a highly irreversible process. As yet no practical method has been devised which can reversibly release the energy stored in fuels on a large scale.

$\eta_{a,c,} < 1$: All reversible cycles have efficiencies less than 1; and all real cycles involve both internal and external irreversibilities.

(ii) *Upper temperature limit of the cycle*

This is determined by the strengths of materials used in making the components, notably the boiler and turbine. It is about 900 K for the steam plant, although the exact figure depends on the life of the plant. Since steam has a critical state of 22.12 MN/m^2 and 647.30 K, the metallurgical limit of the metal cannot be reached without superheating.

As increases in temperature are accompanied by increases in pressure, a compromise must be reached between low operating cost and low capital cost.

(iii) *Lower temperature limit of the cycle*

This is governed by two factors;

  (a) the cheapest available temperature of the heat sink; this is usually the atmosphere or river water having an average temperature of 290 K, say;
  (b) the temperature difference required for the transfer of heat; a minimum difference of 10 K is needed to keep condenser to reasonable size.

Thus, the lowest operational temperature is about 300 K. This corresponds to a saturation pressure of about 3.6 kN/m^2 or 0.036 bar.

Criteria of performance

(i) *The ideal cycle or thermal efficiency*

This is the efficiency of an internally reversible cycle. The maximum ideal cycle efficiency is the Carnot efficiency which is only achieved by both internally and externally reversible cycles.

All actual cycles are less efficient than the corresponding ideal cycles. The efficiency ratio is defined as

$$\text{efficiency ratio} = \frac{\text{actual cycle efficiency}}{\text{ideal cycle efficiency}} \tag{6.8}$$

(ii) *The work ratio, r_w*

This is defined by

$$r_w = \frac{\text{net work}}{\text{positive work}} \tag{6.9}$$

Both high thermal efficiency and high work ratio are required to render high overall efficiency. The necessity for a high work ratio may not be immediately obvious, but it is nevertheless required because a low work ratio implies that the positive and negative works of the cycle are about, equal making the cycle efficiency highly vulnerable to internal irreversibilities.

(iii) *Specific steam consumption*

This is defined as the mass of steam required to produce a unit of power output.

It is given by

$$\text{s.s.c.} = \frac{3600}{w} \text{ kg/kW h} \qquad (6.10)$$

where w is the net specific work output in kJ/kg of steam. This decides the size of the plant.

The Carnot cycle

Since the working fluid is a condensable vapour, the two isothermal processes can easily be attained by heating and cooling the wet vapour at constant pressure. Thus, the T–s diagram appears as shown in Figure 6.8

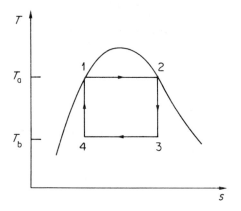

Figure 6.8

During process 1–2 saturated water at state 1 is evaporated in the boiler to form saturated steam at state 2; the heat added is

$$q_{12} = \Delta h = h_2 - h_1$$

Process 2–3 involves the isentropical expansion of saturated steam at state 2 to state 3 corresponding to the temperature of the heat sink; the expansion work is

$$w_{23} = - \Delta h = h_2 - h_3$$

In process 3–4, wet steam at state 3 is passed through the condenser. It is partially condensed at constant pressure. Condensation stops at state 4 where $s_4 = s_1$. The heat rejected is

$$q_{34} = \Delta h = h_4 - h_3$$

Lastly, during process 4–1, wet steam at state 4 is compressed isentropically to state 1, where it becomes fully saturated. The compression work is

$$w_{41} = - \Delta h = h_4 - h_1$$

The net work is therefore

$$w = w_{23} + w_{41} = h_2 - h_3 + h_4 - h_1$$

$$= h_2 - h_1 + h_4 - h_3 = q_{12} + q_{34}$$

The change in entropy during the heating process is

$$s_2 - s_1 = \frac{q_{12}}{T_a}$$

and the change in entropy during the cooling process is

$$s_4 - s_3 = \frac{q_{34}}{T_b}$$

But the total change in entropy for the cycle is zero, therefore

$$s_2 - s_1 + s_4 - s_3 = \frac{q_{12}}{T_a} + \frac{q_{34}}{T_b} = 0$$

giving

$$\frac{q_{12}}{q_{34}} = - \frac{T_a}{T_b}$$

By definition, the thermal efficiency is given by

$$\eta_{th} = \frac{w}{q_{12}} = \frac{q_{12} + q_{34}}{q_{12}} = 1 - \frac{T_b}{T_a} \qquad (6.11)$$

and the work ratio is

$$r_w = \frac{w}{w_{23}} \qquad (6.12)$$

Impracticability of Carnot cycle

Though the Carnot cycle has the highest possible theoretical thermal efficiency, it is difficult to realize in practice for the following reasons.

(1) It has a low work ratio.
(2) Practical difficulties are associated with compression because it is difficult

to control the condensation process at 4; also 4–1 involves the compression of a wet vapour for which process no efficient method has been devised, since the liquid tends to separate out on compression and the compressor has to handle a non-homogeneous substance. Also, the volume of fluid handled is high since the pressure is low, i.e. the specific volume is high. This requires a large compressor.

(3) The steam is wet throughout the expansion. This leads to excessive wall condensation in a reciprocating engine and to erosion of the blades in a turbine.

The Rankine cycle

This is an attempt to avoid the practical difficulties associated with the compression and condensation processes of the Carnot cycle. In this cycle, steam is totally condensed to liquid so that a pump can be used to keep up the circulation. When both the expansion of the steam and the pumping of the condensate occur reversibly, the cycle is called the ideal Rankine cycle.

For two engines operating between the same temperatures

$$\eta_{\text{Rankine}} < \eta_{\text{Carnot}}$$

since not all the heat is supplied at the upper temperature, i.e. the Rankine cycle suffers from external irreversibilities. But the net specific work output is greater than the corresponding Carnot cycle. Hence, the steam consumption is less and the work ratio greater. The Rankine cycle is shown in Figure 6.9. The expansion work is

$$w_{23} = h_2 - h_3$$

Pumping work is

$$w_{45} = h_4 - h_5 = v_f (p_4 - p_5)$$

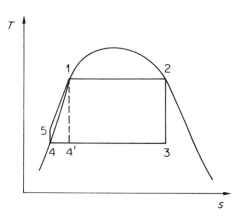

Figure 6.9

The net work is therefore

$$w = w_{23} + w_{45}$$

The heat supplied is

$$q_{52} = h_2 - h_5$$

hence, the thermal efficiency is

$$\eta_{th} = \frac{w}{q_{52}} \qquad (6.13)$$

and the work ratio is

$$r_w = \frac{w}{w_{23}} \qquad (6.14)$$

When internal irreversibilities are to be taken into account, the work processes will tend towards the right of the isentropic processes as outlined in Section 6.1. The isentropic efficiencies can then be used to compute the actual work interaction. The actual processes would reduce both cycle efficiency and work ratio of the cycle.

Methods for improving cycle efficiencies

(i) *Decreasing the exhaust pressure*

This is equivalent to decreasing the lower operating temperature of the engine. The lowest operating temperature corresponds to that of readily available cooling water.

(ii) *Increasing the boiler pressure*

This is equivalent to increasing the upper operating temperature. The highest possible increase is governed by the metallurgical limit of materials used for the construction of the plant.

For the Carnot cycle, an increase in boiler pressure increases the efficiency continuously, though as the pressure increases the gain becomes progressively smaller, with marginal returns. For the Rankine cycle, the efficiency actually reaches a maximum around 15 MN/m² or 150 bars. In fact, the efficiency curve is quite flat over the range 10–20 MN/m².

(iii) *Superheating the Rankine cycle: superheated Rankine cycle*

This increases the cycle efficiency compared to the Rankine cycle operating at the same boiler pressure since now heat is supplied over a higher average temperature. However, the efficiency ratio referred to a Carnot cycle operating with the same increased temperature heat source is in fact lower. This will not unduly worry us since the Carnot cycle is impossible to operate. The work ratio remains essentially constant whether the Rankine cycle is superheated or not. However, the specific steam consumption is markedly reduced. The added complexity of a superheater is therefore amply compensated by a reduction in size of the other components.

Increasing the superheat increases the cycle efficiency continuously; the highest superheat temperature is therefore limited only by the metallurgical limits of the components.

A further advantage of superheating is the increase in the quality of the steam after expansion. This becomes of greater importance the higher the boiler pressures. Wet steam leads to excessive wall condensation in a reciprocating engine, or erosion of the turbine blades as well as reduction of the turbine isentropic efficiency. In practice, the quality of the steam at the turbine exit is not allowed to fall below 0.9.

The T–s diagram of a superheated Rankine cycle is shown in Figure 6.10.

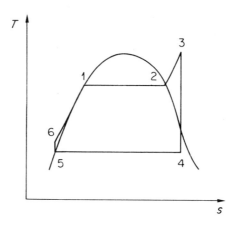

Figure 6.10

(iv) *Reheat cycle*

In the Reheat cycle expansion takes place in several stages. Reheating is assumed to take place at constant pressure and usually to the original superheat temperature. Reheating makes very little difference to cycle efficiency, which is generally lower than that without reheating. The main advantage offered by reheating is to improve the quality of steam in the turbine. It also reduces appreciably the specific steam consumption. This can be understood from consideration of the area bounded which represents the specific work output.

A Rankine cycle with óne stage superheat and one stage reheat is shown in Figure 6.11.

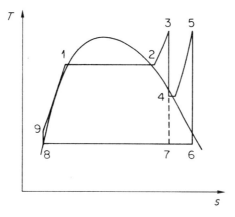

Figure 6.11

(v) *Regenerative cycle*

The efficiency of the Rankine cycle is smaller than that of the Carnot cycle because all the heat is not supplied at the upper temperature. Some heat is added while the liquid goes from T_5 to T_1. If some means could be found to transfer this heat reversibly from the working fluid in another part of the cycle, then all the heat supplied from an external source would be transferred at the upper temperature and Carnot efficiency would then be achieved.

A cycle in which this principle is used to raise the efficiency of the Rankine cycle is called a regenerative cycle.

One method which is theoretically possible is to pass the feedwater around the turbine casing at the point where the expanding steam has a temperature slightly greater than T_6 (see Figure 6.12 for this state). The water flows in the

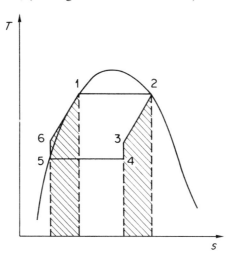

Figure 6.12

opposite direction to the steam and leaves the turbine at a temperature slightly smaller T_1 before entering the boiler.

At all points heat is transferred at a small temperature difference and the process tends to a reversible one.

Most expansion in the turbine is not adiabatic. Since the heat rejected in 2–3 equals that absorbed in 6–1, the shaded areas in the T-s diagram are equal; thus

$$\eta_{cycle} = \eta_{Carnot}$$

The above cycle is impracticable for two reasons.

(1) It is impossible to design a turbine operating efficiently as a turbine and a heat exchanger at the same time.
(2) The quality of the expanded steam is too low.

As an alternative, the following practical regenerative cycle, using feedwater heater, has been adopted. The T-s diagram for a simple regenerative cycle using one feedwater heater is shown in Figure 6.13.

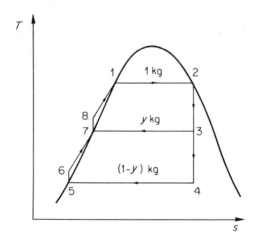

Figure 6.13

Let us examine the operation of this cycle more closely by considering the processes that 1 kg of steam goes through. It is expanded at state 2 to an intermediate state 3 at which point y kg is led off to mix with the feedwater, which comprises of the remaining l-y kg of steam after being fully expanded to the condenser pressure. Before mixing with the bled steam, this 1−y kg of steam at state 4 is first condensed to state 5 and then pumped into the feed heater at state 6.

In this cycle, the average temperature at which heat is added from the external source is higher than that of the Rankine cycle; it is therefore, more efficient.

The cycle we have considered above requires a feed pump after each feed heater, since the various fluid streams are allowed to mix with each other. It is known as an open feed heating process. The number of pumps required therefore equals the number of heaters plus one. For this reason, it is not often used in practice.

A variation of the above scheme is obtained by preventing the bled steam from mixing with the feedwater. After passing through the feed heater the steam at a higher pressure is throttled to the condenser pressure to blend with the fully expanded stream. Only one feed pump is now required. This scheme is known as closed feed heating and is more widely used.

The T–s diagram for a vapour cycle with one stage closed system feed heating is shown in Figure 6.14.

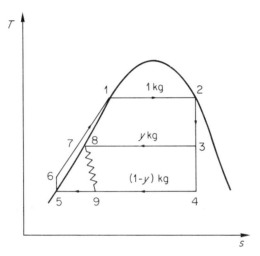

Figure 6.14

(vi) *Binary cycle*

As the greatest portion of heat imput in the Rankine cycle is added during evaporation of the liquid (the amounts added to the feedwater and during superheating is relatively small), a major improvement in the cycle can be achieved if the evaporator can be used at a temperature close to the metallurgical limit of the turbine. The limitation imposed on the efficiency of a steam power plant is therefore due to the properties of the working fluid–steam has a relatively low T_c (647.3 K) and high p_c (22.1 MN/m^2). It is therefore desirable to have fluid whose critical temperature is well above the metallurgical limit of 900 K and whose vapour pressure remains moderate at 900 K. Mercury satisfies these requirements fully (at T = 900 K its saturation pressure is about 3.0 MN/m^2 only). However, mercury has the disadvantage that at usual heat rejection temperatures its vapour pressure is exceedingly low and its specific volume enormous. Also, its low latent heat of vaporization (about 1/8 that of steam) implies a high specific mercury consumption.

In order to operate over the whole temperature range from 300–900 K, a dual cycle using both mercury and steam has been proposed. Such a scheme is known as a binary cycle.

In the binary cycle, the mercury condenser serves as the steam boiler. The *T–s* diagram for such a combination with no reheat or regenerative heating but with steam superheating is shown in Figure 6.15.

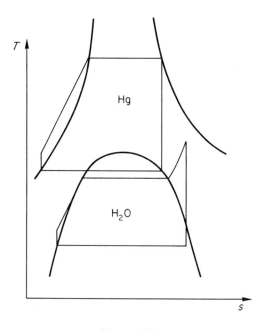

Figure 6.15

Binary cycles can give a significant improvement on the thermal efficiency of the plant. However, because mercury is highly poisonous, attacks many common metals, does not wet the boiler and condenser surfaces (giving rise to difficulty in design of these components) and is expensive, it is not found in ordinary power plants.

Ideal working fluid

Apart from the Carnot cycle, which is not of much practical importance, all the cycles discussed above have efficiencies and specific consumptions dependent on the properties of the working fluid. Although for reasons of cost and chemical stability steam is almost invariably used in vapour cycles, its behaviour is far from ideal. The characteristics of an ideal working fluid are briefly discussed below. A *T–s* diagram for such a fluid is shown in Figure 6.16; its main features are as follows.

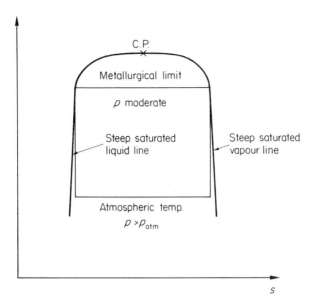

Figure 6.16

(a) The critical temperature is well above the metallurgical limit, so that most of the heat input can occur at the highest practical temperature to render superheating unnecessary.

(b) The saturation pressure at the metallurgical limit is moderate so that capital and maintenance costs can be kept low.

(c) The specific heat of the liquid is small, giving a steep saturated liquid line. Thus the heat required to bring the cold liquid to the boiling temperature is small, enabling most of the heat to be added at the upper temperature.

(d) The latent heat of vaporization is high, resulting in low specific consumption and hence a small plant for a given power output.

(e) The saturated vapour line is steep so that the quality of the expanded fluid is greater than 0.9 without recourse to superheating and reheating, which increase the complexity of the plant.

(f) The saturation pressure at the lowest available temperature of 300 K is slightly higher than atmospheric so that no vacuum is needed in the condenser.

In addition, it is desirable that the fluid be cheap, chemically stable, non-toxic and non-corrosive.

6.3 GAS POWER CYCLES

In the bulk of gas power plants, the combustion of the fuel is usually performed within the engine itself. For this reason, such an engine has been called an internal combustion engine (ICE). The working fluid therefore consists of air and fuel vapour before combustion, and the products of combustion afterwards. ICEs invariably operate as 'open' cycles, since a fresh charge has to be admitted every time for combustion to be carried out. The only exception to this mode of operation is the closed cycle gas turbine where the working fluid is recirculated with heat input from a heater.

In the simplest approach to gas cycle analysis, it is assumed that the specific heats of the fluid are those of air, that the process of combustion does not affect their values, that they are independent of temperature and that the weight of the charge is unaffected by the injected fuel. The efficiency thus obtained is referred to as the 'air standard efficiency'.

The following air standard cycles form the basis for analysing ICE cycles;

(a) the Otto cycle for petrol or spark-ignition (SI) engines;
(b) the Diesel cycle—except for large marine engines, very few engines are left today which employ this cycle;
(c) the mixed or dual cycle—the ideal cycle for compression–ignition (CI) engines, which are commonly referred to as 'Diesel' engines; in actual fact, many of the modern CI engines approximate as closely to the Otto cycle as petrol or gas engines;
(d) the Joule (or Brayton) cycle for gas turbines.

Criteria of performance

(i) *Air standard efficiency*

This corresponds to the thermal efficiency for vapour cycles and is similarly defined, i.e.,

$$\eta_{th} = \frac{\text{net work/cycle}}{\text{heat input/cycle}} \tag{6.15}$$

(ii) *Indicated thermal efficiency*

$$\eta_i = \frac{\text{indicated work/cycle}}{\text{heating value of fuel supplied/cycle}} \tag{6.16}$$

This gives the actual cycle efficiency before friction losses are taken into account.

From (i) and (ii), an actual or indicated efficiency ratio can be defined as

$$\text{efficiency ratio} = \frac{\eta_i}{\eta_{th}} \tag{6.17}$$

(iii) *Specific fuel consumption*

This gives a measure of the economy of operating the engine. It may be expressed on the basis of either the brake power output or the indicated power output.

As mechanical friction is of little interest here, only the indicated specific fuel consumption will be adopted. This can be defined as

$$\text{Isfc} = \frac{3600}{(\text{H.V. of fuel})\,\eta_\text{I}} \quad \text{kg/kW h} \tag{6.18}$$

where H.V., the heating value of the fuel is in units of kJ/kg.

Other bases used for comparing piston engines are discussed below.

(iv) *Mean effective pressure* (*mep*)

It has been seen in steady flow vapour cycles that the work ratio gives an indication of the size of the plant per unit power output. This criterion, although it can be computed for theoretical cycles, is rather meaningless for experimental purposes, since it is not possible to isolate the positive and negative works in the cycle as all the processes are carried out in a single component. An alternative criterion known as the mean effective pressure is used instead. This is defined as the height of a rectangle on the p–v diagram having the same length and area as the cycle. Since the area measures the work output per cycle, it can therefore be expressed as

$$\text{mep} = p_\text{m} = \frac{\text{work output/cycle}}{\text{swept volume}} \tag{6.19}$$

From this definition, it is evident that a cycle with a large p_m will produce a large work output per unit swept volume and hence the engine is smaller for a given work output.

The above definition can easily be converted to an experimentally measurable quantity by using the indicated work for the numerator and the actual piston swept volume for the denominator. In this form, it is known as the indicated mean effective pressure, (Imep), where

$$\text{Imep} = \frac{\text{indicated work}}{\text{piston cross-section} \times \text{piston stroke}} \tag{6.20}$$

When all quantities are measured in SI units, the Imep is in N/m^2.

(v) *Indicated power output*

This represents the power delivered to the piston from the working medium. It is defined as

$$\text{I.p.} = \text{indicated work} \times \text{engine speed}$$

$$= p_\text{m,i} \, LAN \quad W \tag{6.21}$$

where L is the piston stroke (m), A is the piston cross-section (m^2) and N is the engine speed in cycles per second; this equals the shaft speed (rev/s) for 2-stroke engines and equals $\frac{1}{2}$ (shaft speed) for 4-stroke engines.

Air-standard Otto cycle

This comprises two isentropic and two isometric processes as shown in Figure 6.17.

The processes undergone by the fluid in the cylinder are;

   1–2: isentropic air compression through a volume ratio $r_c = v_1/v_2$, known as the compression ratio;
   2–3: heat addition q_a at constant volume at top-dead centre, TDC;
   3–4: isentropic air expansion to original volume at bottom dead centre, BDC.
   4–1: heat rejection q_b at constant volume at BDC.

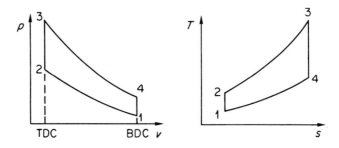

Figure 6.17

Analysis

$$q_a = u_3 - u_2$$

$$q_b = u_1 - u_4$$

$$w = q_a + q_b = (u_3 - u_2) - (u_4 - u_1)$$

therefore

$$\eta_{th} = \frac{w}{q_a} = 1 - \frac{u_4 - u_1}{u_3 - u_2}$$

If the working fluid, air, is a perfect gas, then

$$\Delta u = c_v \, \Delta T$$

and

$$\eta_{th} = 1 - \frac{T_4 - T_1}{T_3 - T_2}$$

For the isentropic processes

$$\frac{T_2}{T_1} = \frac{T_3}{T_4} = r_c^{\gamma-1} \qquad (6.22)$$

therefore

$$\eta_{th} = 1 - \frac{1}{r_c^{\gamma-1}}$$

Also

$$p_m = \frac{w}{v_1 - v_2} = \frac{c_v(T_3 - T_2 - T_4 + T_1)}{v_1 - v_2}$$

$$= \frac{c_v(T_3 - T_3 r_c^{1-\gamma} - T_2 + T_2 r_c^{1-\gamma})}{v_2(r_c - 1)}$$

$$= \frac{c_v}{r_c - 1}\left(\frac{T_3}{v_3} - \frac{T_2}{v_2}\right)(1 - r_c^{1-\gamma})$$

(because $v_2 = v_3$)

$$= \frac{c_v}{R(r_c - 1)}(p_3 - p_2)\left(1 - \frac{1}{r_c^{\gamma-1}}\right)$$

(because $pv = RT$)
But

$$c_p - c_v = R \quad \text{or} \quad \frac{c_v}{R} = \frac{1}{\gamma - 1}$$

and

$$p_2 = p_1\left(\frac{v_1}{v_2}\right)^\gamma = p_1 r_c^\gamma$$

hence

$$p_m = \frac{(p_3 - p_1 r_c^\gamma)\left(1 - \frac{1}{r_c^{\gamma-1}}\right)}{(\gamma - 1)(r_c - 1)} \tag{6.23}$$

Internal combustion engines

It has been mentioned in the introduction to the present section that the Otto cycle is in fact a close approximation for almost all high speed reciprocating engines presently in use, so the discussion will digress somewhat to give a brief description of the general performance of internal cumbustion engines before proceeding to more gas cycle analysis. The two main results of our foregoing analysis are obviously the thermal efficiency and the mean effective pressure. Of these, the efficiency has been found to be a function of the compression ratio only; the higher the ratio the higher efficiency. This is perhaps the most important result deduced. The following table shows how η_{th} varies with respect to r_c.

r_c	η_{th}		
2	0.240		
3	0.353		
4	0.423		
5	0.471	0.34	Values of efficiency
6	0.508	0.37	obtained assuming
7	0.537	0.395	variable specific heat
8	0.561	0.415	and dissociation of
10	0.598	0.45	combustion gases
12	0.626		
14	0.648		
16	0.666		
18	0.682		
20	0.695		

The results of η_{th} computed with the air-standard cycle formula are obviously too high for practical engines; values of efficiencies obtained by assuming variable specific heat and dissociation of combustion gases have also been given for comparison. These latter values are much more reasonable. However, because of the complexity of the calculations involved, air-standard cycles are still of immense value in cycle analysis in helping us to establish qualitatively the influence of the important variables on performance.

Present day petrol engines have compression ratios from 6–10. It is necessary to use premium grade petrols in engines with high compression ratios to avoid pre-ignition or detonation of the pre-combustion mixture in the engine cylinder. Detonation leads to loss of power as well as damage to the internal engine surfaces.

The range of compression ratio commonly employed in oil or diesel engines is from 12–18. Thus the efficiencies of oil engines tend to be higher than those of petrol engines. Higher compression ratios are used in oil engines because it is necessary to compress the air to a high enough temperature, so that ignition of the fuel oil will take place when it is injected into the cylinder in the absence of other assistance, such as a spark in petrol engines. Because of the higher pressures attained in oil engines, they are generally heavier in construction.

Diesel cycle

This is shown in Figure 6.18. It comprises two isentropic, one isobaric and one isometric process.

1–2: air is compressed isentropically through a compression ratio $r_c = v_1/v_2$.

2–3: heat input q_a occurs at a constant pressure while the air is expanding from v_2 to v_3. The ratio $r_{co} = v_3/v_2$ is known as the cut-off ratio.

3–4: air is expanded isentropically to its original volume v_1.

4–1: heat rejection q_b occurs at constant volume.

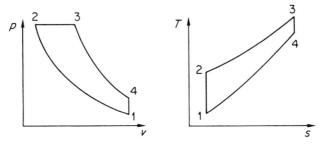

Figure 6.18

Analysis

$$w = q_a + q_b = (h_3 - h_2) + (u_1 - u_4)$$

Therefore

$$\eta_{th} = \frac{w}{q_a} = 1 - \frac{u_4 - u_1}{h_3 - h_2}$$

but

$$\Delta u = c_v \Delta T \text{ and } \Delta h = c_p \Delta T$$

hence

$$\eta_{th} = 1 - \frac{c_v (T_4 - T_1)}{c_p (T_3 - T_2)} = 1 - \frac{1}{\gamma} \left(\frac{T_4 - T_1}{T_3 - T_2} \right)$$

Using the relations

$$r_c = \frac{v_1}{v_2} \text{ and } r_{co} = \frac{v_3}{v_2}$$

$$\eta_{th} = 1 - \frac{1}{r_c^{\gamma-1}} \left\{ \frac{r_{co}^\gamma - 1}{\gamma (r_{co} - 1)} \right\} \tag{6.24}$$

With some more involved manipulations, the mean effective pressure can be shown to be given by

$$p_m = p_1 \left\{ \frac{\gamma r_c^\gamma (r_{co} - 1) - r_c (r_{co}^\gamma - 1)}{(\gamma - 1) (r_c - 1)} \right\} \tag{6.25}$$

Comparing the efficiencies for the Otto and Diesel cycles shows that the two differ only by the term in brackets. This term is a function of load ratio and is always greater than 1 for finite loads. It follows that at a given r_c the Otto cycle has a higher efficiency, work output and mep than the Diesel cycle. At low-load conditions where $r_{co} \to 1$, $\eta_{Diesel} \to \eta_{Otto}$ or actually exceeds it, because of the absence of throttling effects. In practice, the high efficiency generally associated with CI engines results from the use of higher r_c. This has been discussed in the previous section.

Dual or mixed cycle

The processes constituting the dual cycle are shown in Figure 6.19. These are;

1–2: isentropic compression;
2–3–4: heat addition q_a partly at constant volume and partly at constant pressure;
4–5: isentropic expansion;
5–1: heat rejection q_b at constant volume.

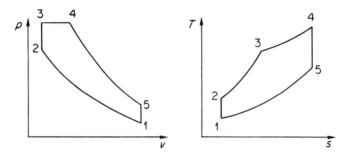

Figure 6.19

Analysis

$$w = q_a + q_b = c_v(T_3 - T_2 + T_1 - T_5) + c_p(T_4 - T_3)$$

therefore

$$\eta_{th} = \frac{w}{q_a} = 1 - \frac{T_5 - T_1}{(T_3 - T_2) + (T_4 - T_3)}$$

Letting

$$r_c = \frac{v_1}{v_2} ; r_{co} = \frac{v_4}{v_3}$$

and

$$r_p = \frac{p_3}{p_2} = \text{constant volume pressure ratio}$$

η_{th} can be written as

$$\eta_{th} = 1 - \frac{1}{r_c^{\gamma-1}} \left\{ \frac{r_p r_{co}^{\gamma} - 1}{(r_p - 1) + \gamma r_p(r_{co} - 1)} \right\} \tag{6.26}$$

Likewise, the mean effective pressure can be found to be

$$p_m = \frac{p_1 \{r_c^\gamma (r_p - 1) + r_c (1 - r_p r_{co}) + \gamma r_p r_{co}^\gamma (r_{co} - 1)\}}{(\gamma - 1)(r_c - 1)} \qquad (6.27)$$

Joule or Brayton cycle

This is the ideal cycle for the operation of the simple gas turbine. Here a steady-flow process is involved; hence, the constant pressure processes. The cycle is shown in Figure 6.20.

The cycle comprises the following processes;

1–2: isentropic compression;
2–3: heat addition q_a at constant pressure;
3–4: isentropic expansion of products of combustion;
4–1: heat rejection q_b at constant pressure.

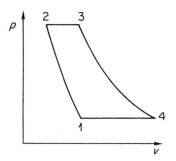
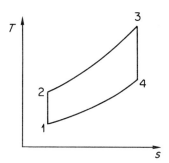

Figure 6.20

Analysis
Under the simplifying assumptions listed earlier in the introduction to the present section, the analysis for both the open–and closed–cycle gas turbines are identical.

$$q_a = c_p (T_3 - T_2) \text{ and } q_b = c_p (T_1 - T_4)$$

Therefore

$$w = c_p (T_3 - T_2 - T_4 + T_1)$$

hence

$$\eta_{th} = 1 - \frac{T_4 - T_1}{T_3 - T_2} = 1 - \frac{1}{r_c^{\gamma - 1}} = 1 - \frac{1}{r_p^{(\gamma - 1)/\gamma}} \qquad (6.28)$$

where r_c is the compression ratio v_1/v_2 and r_p is the pressure ratio p_2/p_1

As gas turbines operate in steady-flow, we refer back to r_w as a performance criterion; this is given by

$$r_w = \frac{w}{w_+} = \frac{T_3 - T_2 - T_4 + T_1}{T_3 - T_4} = 1 - \frac{T_2 - T_1}{T_3 - T_4} = 1 - \frac{T_1}{T_3} r_p^{\gamma - 1/\gamma} \quad (6.29)$$

Like the Otto cycle, the efficiency of the ideal cycle for a simple gas turbine is a function of the compression ratio, which in the present case, is also a function of the pressure ratio only. The work ratio, which measures the cycle's susceptibility to irreversibilities, however, depends also on the minimum and maximum temperatures of the turbine. In practice, T_1 = atmospheric temperature = 290 K, say; and T_3 = metallurgical limit of the turbine. This depends on the life of the plant and the heat resistant alloy of which the plant is made. Let us take this to be 1000 K say. Having fixed T_3 and T_1, the maximum pressure ratio possible is therefore given by

$$r_{p,\max} = \left(\frac{T_3}{T_1}\right)^{\gamma/(\gamma - 1)} \quad (6.30)$$

The net work is therefore

$$w = c_p(T_3 - T_2) + c_p(T_1 - T_4)$$

$$= c_p T_3 \left(1 - \frac{1}{r_p^{(\gamma-1)/\gamma}}\right) - c_p T_1 (r_p^{(\gamma-1)/\gamma} - 1)$$

As the temperatures T_1 and T_3 have been fixed, the maximum w can be obtained by equating $dw/dr_p = 0$. This gives

$$r_p = \left(\frac{T_3}{T_1}\right)^{\gamma/2(\gamma - 1)} \quad \text{or} \quad r_p = \sqrt{(r_{p,\max})} \quad (6.31)$$

With values for T_3 and T_1 quoted above, the optimum r_p is 8.8. At this r_p the size of the plant is minimal. The corresponding work ratio is

$$r_w = 1 - \sqrt{\frac{T_1}{T_3}} = 0.452$$

Thus over one half of the turbine work is consumed in the compressor.

Improving gas turbine performance

We have seen earlier from vapour cycles that the efficiency of a steady-flow cycle can be increased by reheat and regenerative heating. The same techniques can be applied to the gas turbine cycle. These are discussed below.

(i) Simple gas turbine cycle with one stage regenerative heating

In this modification, the hot exhaust gases are used to heat the air intake before it passes into the combustion chamber or heater. The *T–s* diagram for a simple gas turbine cycle with ideal regenerator is shown in Figure 6.21. The effect of regenerative heating is to reduce the heat supply, which now becomes

$$q_a = h_3 - h_x$$

The heat rejected is also reduced to

$$q_b = h_y - h_1$$

The compressor and turbine works are respectively

$$w_C = h_2 - h_1 \quad \text{and} \quad w_T = h_3 - h_4$$

Therefore

$$\eta_{th} = \frac{w}{q_a} = \frac{(h_3 - h_4) - (h_2 - h_1)}{h_3 - h_x} = \frac{(T_3 - T_4) - (T_2 - T_1)}{T_3 - T_4}$$

$$= 1 - \frac{T_1}{T_3} r_p^{(\gamma-1)/\gamma} \tag{6.32}$$

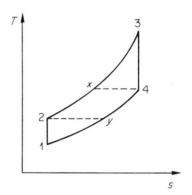

Figure 6.21

Since the work terms remain unchanged, it is seen that

$$\eta_{th} = r_w \tag{6.33}$$

In this case, the efficiency becomes also temperature dependent and the effect of the pressure ratio (r_p) is reversed when compared to the simple cycle. The lower the pressure ratio the higher the efficiency, the maximum value being $(T_3 - T_1)/$

T_3 which occurs when $r_p = 1$. At this point, though both r_w and η_{th} are high, the cycle is impractical because the actual work output is very small—tending to zero. When both the efficiencies from the simple and regenerative cycles are equal, i.e. when regenerative heating becomes superfluous

$$r_p = \left(\frac{T_3}{T_1}\right)^{\gamma/2(\gamma-1)} \tag{6.34}$$

which is the optimum r_p for maximum work output. Clearly, for regenerative heating to be advantageous, $r_p < r_{p,max}$.

(ii) One stage reheating, one stage regenerative heating and one stage intercooling

When a gas is being compressed, its temperature rises rapidly. Cooling of this hot gas at constant pressure reduces its volume and acts therefore to reduce the compressor work, thus augmenting cycle performance.

In the present modification, the working fluid is being compressed in two stages with an intercooler between these stages. Expansion also takes place in two stages with reheating between the stages. The T–s diagram is as shown in Figure 6.22.

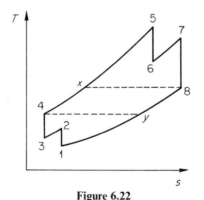

Figure 6.22

The heat supplied from the external source is now supplied in the heater and reheater; thus

$$q_a = (h_5 - h_x) + (h_7 - h_6)$$

and the heat rejected is

$$q_b = (h_y - h_1) + (h_2 - h_3)$$

If the working substance is a perfect gas, and an ideal regenerator is assumed, both the thermal efficiency and work ratio can be shown to be

$$1 - \frac{T_4 - T_1 + T_2 - T_3}{T_5 - T_8 + T_7 - T_6} \tag{6.35}$$

Cycles having Carnot efficiency

These are reversible cycles operating between two isotherms. The Carnot cycle, as we have seen, is closed by two other isentropics. Two other cycles of interests are the Stirling and Ericsson cycles. The former is closed by two isometrics and the latter by two isobars.

In order that the Stirling and Ericsson cycles should approach the Carnot efficiency, the heat rejected by the hot gases during expansion must be transferred to the cold gases being compressed. With a perfect regenerator, their efficiencies will be equal to that of the Carnot cycle, which is $(T_a - T_b)/T_a$ where subscripts a and b refer to the temperatures of the heat source and sink respectively.

The mean effective pressure of the Carnot cycle is given by

$$p_m = \frac{p_2 (T_a - T_b) \ln \dfrac{v_3}{v_2}}{\left\{\left(\dfrac{T_a}{T_b}\right)^{1/(\gamma-1)} \left(\dfrac{v_3}{v_2}\right) - 1\right\} T_a} \tag{6.36}$$

For states 2 and 3, refer to Figure 6.23.

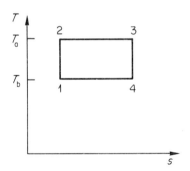

Figure 6.23 Carnot cycle

The mean effective pressures for the Stirling and Ericsson cycles are respectively

$$p_m = \frac{r_c p_1 (T_a - T_b) \ln r_c}{T_b (r_c - 1)} \tag{6.37}$$

$$p_m = \frac{r_c p_1 (T_a - T_b) \ln r_c}{r_c T_a - T_b} \tag{6.38}$$

where

$$r_c = \text{isothermal compression ratio}$$

$$= v_4/v_1$$

Refer to Figure 6.24 for states 4 and 1. On the T–s diagram, the Stirling and Ericsson cycles look quite similar, although the isometrics have steeper gradients than the isobars.

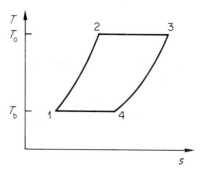

Figure 6.24 Stirling or Ericsson cycle

6.4 REFRIGERATION CYCLES

In a broad sense, refrigeration concerns the production and maintenance of temperatures below that of the surrounding atmosphere. The net result is the extraction of heat from a low temperature region and its rejection at a higher temperature. Thus, a refrigeration cycle functions simultaneously as a heat pump.

Low temperatures can be obtained by the following methods;

(1) allowing a phase change to take place in such a way as to extract heat;
(2) expanding a compressed gas or vapour so that it does external work;
(3) the Joule–Kelvin expansion of a gas;
(4) making use of the Peltier effect, i.e. passing an electric current through a bimetallic junction;
(5) demagnetization of a solid;
(6) gaseous desorption.

Practical refrigeration cycles therefore require that one or more of the above processes should be operated in a continuous manner so that the low temperature produced can be maintained. Two general types of refrigeration cycles are in common use, both depending on the attainment of a cold region by the vaporization of a refrigerant from an evaporator at low temperature and pressure. The first makes use of mechanical work to actuate the cycle, and is called compression refrigeration. In the second, known as absorption refrigeration, heat is used. In compression refrigeration, work is done to compress the vapour leaving the evaporator so that the heat absorbed at the low temperature of the evaporator can be rejected at the temperature level of the condenser. In absorption refrigeration, heat absorbed at a low temperature and pressure level is rejected at an intermediate temperature and higher pressure level after its temperature has been increased by heat addition from a high temperature source such as the combustion of a fuel.

Performance criteria

(i) *Coefficient of performance*

The performance index of engineering systems are always defined as the ratio of what is desired to what is paid for. For a refrigeration cycle, this takes the form

$$COP = \frac{\text{refrigeration effect/cycle}}{\text{net work or heat input/cycle}} \qquad (6.39)$$

Since this ratio is generally larger than 1, it is not called an efficiency. Instead, it is known as the coefficient of performance, abbreviated as COP.

(ii) *The refrigeration effect*

This is the heat absorbed per kilogram of refrigerant. It provides a basis for comparing the sizes of plants required for a given duty (heat extracted per unit time). For practical purposes, a comparison of refrigeration effects per unit *volume* of refrigerant is more useful, because the displacement of the compressor and the size of the plant in general depends mainly on this parameter.

(iii) *The net work or heat input*

This provides an estimate of the operating cost of the plant.

Ideal compression refrigeration cycles

(i) *Vapour-compression cycle: Reversed Carnot cycle*

With a condensable fluid, the isothermal processes of a reversed Carnot cycle can easily be carried out by having a refrigerant that condenses during the heat rejection process and boils during the heat absorption process. The cycle comprises the following processes;

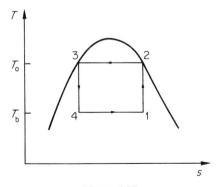

Figure 6.25

114

1–2: reversible adiabatic compression;
2–3: reversible isothermal rejection of heat at high temperature;
3–4: reversible adiabatic expansion;
4–1: reversible isothermal absorption of heat to low temperature surroundings.

Analysis

$$q_b = T_b (s_1 - s_4)$$

$$q_a = T_a (s_3 - s_2) = - T_a (s_1 - s_4)$$

Therefore

$$w = q_a + q_b = - (T_a - T_b) (s_1 - s_4)$$

thus

$$\text{COP} = \frac{q_b}{w} = \frac{T_b}{T_a - T_b} \qquad (6.40)$$

Hence, for high COP, we require the cycle to operate with high T_b and low T_a.

(ii) *Gas compression cycle*: *Reversed Joule cycle*

When gases are used as refrigerants, the ideal cycle is the reversed Joule or
Brayton cycle because now the cooling and heating processes are more practically arranged to take place at constant pressure in steady flow. Consequently,
a refrigerator using a gas as working fluid is less efficient than the one using a
vapour. Furthermore, it is much bulkier than a vapour plant of the same duty
because a gas requires a relatively larger surface area for a given heat transfer.
Also, for a given operational temperature range, its operating temperature
range is much wider, as can be seen from Figure 6.26. For these reasons, gases

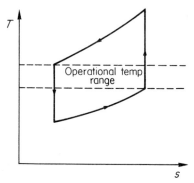

Figure 6.26

are never used nowadays as refrigerants, with the exception that the air cycle has found commercial acceptance in aircraft air-conditioning units.

However, gas cycles are still of great importance with respect to their applications to the liquefaction of gases.

Practical compression refrigeration cycles

Contrary to heat engines, refrigeration machines are usually much smaller devices as their product is of more limited application. The net work employed in a refrigeration cycle is generally small and the positive work is even smaller. It is for this reason that the positive work is seldom large enough to justify the cost of the equipment necessary to obtain it. In practical cycles, the expansion is therefore simply achieved by means of a throttle valve. Thus the end states of the expansion process lie on a constant enthalpy line instead of an isentropic. The throttling process introduces an irreversibility which makes the cycle irreversible as a whole. For this reason, we shall call practical cycles theoretical cycles rather than ideal cycles.

A practical cycle differs in two more aspects from the reversed Carnot cycle. Firstly, the compression is usually carried out in the superheat region, for the same reason as discussed in connection with the Carnot cycle. Secondly, the condensed liquid is often subcooled before entering the throttle valve. Incorporating these modifications into the reversed Carnot cycle, we have the following theoretical cycle (Figure 6.27) where the T–s diagram is shown together with the p–h diagram. This latter form is more useful for refrigeration calculations,

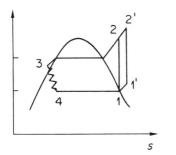 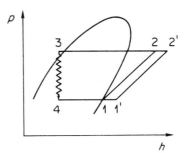

Figure 6.27

since all pertinent quantities can be read off directly from commercially available p–h charts once the cycle has been sketched on it.

The processes involved in the theoretical cycle are;

1–2: reversible adiabatic compression of saturated or superheated vapour;
2–3: reversible rejection of heat at constant pressure;
3–4: throttling of liquid (saturated or subcooled) to evaporator pressure;
4–1: reversible addition of heat at constant pressure in evaporator.

116

Analysis

Refrigeration effect

$$q_{41} = h_1 - h_4$$

net work

$$w = h_2 - h_1$$

Therefore

$$COP = \frac{q_{41}}{w} = \frac{h_1 - h_4}{h_2 - h_1} \qquad (6.41)$$

Ideal absorption refrigeration system

The most widely used absorption refrigeration cycle uses ammonia as the refrigerant and water as the absorbent. A diagrammatic sketch of the machine is shown in Figure 6.28.

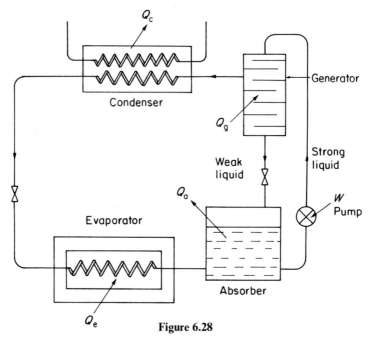

Figure 6.28

This operates on the same cycle as the ordinary compression machine, except that for compression of the vapour a group of 3 processes is substituted for the compressor. These are;

(1) formation of a solution by dissolving the vapour in water;
(2) compression of solution to the higher pressure of the cycle by pumping;
(3) generation of vapour from the solution by heating.

For the ideal system, all interactions with the system are assumed to be reversible. Thus, the generator temperature equals that of the heat source, and the condenser and absorber temperatures are that of the surroundings. Furthermore, it is assumed that the pump work is obtained from a reversible engine operating between the temperatures of the generator, T_g, and the surroundings, T_s, and that this drives a pump with no losses. Under these assumptions, the absorption machine can be reduced to the equivalent black box representation shown in Figure 6.29.

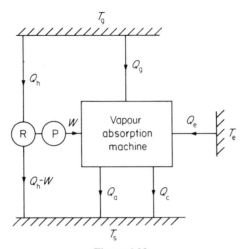

Figure 6.29

Since the machine is working reversibly, the net entropy change of the universe during steady state operation is zero. Thus

$$\Delta S = \frac{Q_c + Q_a + Q_h - W}{T_s} - \frac{Q_g + Q_h}{T_g} + \frac{Q_e}{T_e} = 0$$

where Q_h is the heat input to the Carnot engine at T_g and $Q_h - W$ is the heat rejected by the Carnot engine at T_s

Also, applying the first law to the control surface enclosing the vapour absorption machine, yields

$$Q_g + W + Q_e = Q_a + Q_c$$

Combining the results from the first and second laws

$$(Q_g + Q_h) \frac{T_g - T_s}{T_g} = Q_e \frac{T_s - T_e}{T_e}$$

Thus, the ideal COP of the vapour absorption machine is

$$COP_{ideal} = \frac{Q_e}{Q_g + Q_h} = \left(1 - \frac{T_s}{T_g}\right)\left(\frac{T_e}{T_s - T_e}\right) \tag{6.42}$$

The COP of an ideal reversible absorption system is therefore the product of two quantities. The first represents the Carnot efficiency of the reversible engine working between the high and ambient temperatures, while the second is the COP of a reversed Carnot refrigeration cycle operating between the cold and ambient temperatures. As such, it can be considered as the equivalent of a heat engine and a compression refrigerator working together to produce the desired cooling effect.

Though the COP of an absorption cycle is seen to be decidedly lower than the compression counterpart, the difference is illusory rather than real. This is because in compression cycles, shaft work is required to operate the compressor, and assuming this to be delivered from a Carnot engine without loss, the combined COP for the engine–refrigerator system becomes that of the absorption machine. Thus, we should be aware of the fact that machines working on different fundamental assumptions cannot be compared on the same basis!

Choice of refrigerants

The factors influencing the choice of a refrigerant are;

(1) Moderate vapour pressure within the operational temperature range in order that (a) the volume of the desired amount of fluid shall not be excessively large and so require large and expensive equipment, and (b) the stresses in the containing vessels shall be moderate;
(2) Freezing point lower than the cold temperature in the cycle;
(3) Non-toxic, non-corrosive and odourless;
(4) Stable, i.e. no reaction with lubricating oils;
(5) Non-inflammable and non-explosive;
(6) High latent heat of vaporization at evaporator temperature;
(7) Good wetting ability, good conductivity and low viscosity;
(8) Economical.

A practical refrigerant must also yield a satisfactory COP. Though all would yield the Carnot COP if operated in a reversed Carnot cycle; irreversibilities, due to throttling and other frictional effects, affect the COP to different degrees with different fluids. For instance, $COP_{Carnot} = 5.74$ for $T_a = 304$ K and $T_b = 258$ K, the temperature range of a household refrigerator; for a throttled-cycle, $COP_{NH_3} = 4.85$, $COP_{H_2O} = 4.1$ and $COP_{CO_2} = 2.56$.

EXAMPLES

Example 6.1 (a) Air initially at a pressure of 1.5 MN/m² and a temperature of 523.15 K expands to a pressure of 0.15 MN/m² in a turbine. The isentropic turbine efficiency is 0.80. Find the work done by the turbine and the final temperature of the steam after the expansion.

(b) If the air after expansion is immediately recompressed to its initial pressure before expansion, what is the work required and what will be the final temperature of the air, if the isentropic compressor efficiency is also 0.80.

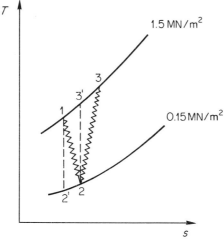

Figure 6.30

(a) The isentropic turbine work, assuming air to be a perfect gas is

$$W_{\text{isen}} = c_p (T_1 - T_2')$$

where

$$\frac{T_2'}{T_1} = \left(\frac{p_2}{p_1}\right)^{(\gamma-1)/\gamma}$$

i.e.

$$T_2' = 523.15 \left(\frac{1}{10}\right)^{0.4/1.4}$$

$$= 523.15/1.93 = 271 \text{ K}$$

therefore

$$W_{\text{isen}} = 1.005 (523.15 - 271)$$

$$= 254 \text{ kJ/kg}$$

thus

$$W_{\text{act}} = \eta_T W_{\text{isen}} = 0.8 \times 254$$

$$= 203 \text{ kJ/kg}$$

Now

$$\eta_T = 0.80 = \frac{T_1 - T_2}{T_1 - T_2'}$$

therefore

$$T_1 - T_2 = 0.8 (523.15 - 271) = 202 \text{ K}$$

hence

$$T_2 = 523.15 - 202 = 321.15 \text{ K}$$

(b)   The isentropic compressor work, assuming air to be a perfect gas is

$$W_{\text{isen}} = c_p (T_3' - T_2)$$

where

$$T_3' = 321.15 \times 1.93 = 620 \text{ K}$$

therefore

$$W_{\text{isen}} = 1.005 (620 - 321.15) = 300 \text{ kJ/kg}$$

thus

$$W_{\text{act}} = \frac{W_{\text{isen}}}{\eta_C} = \frac{300}{0.8} = 375 \text{ kJ/kg}$$

Now

$$\eta_C = 0.80 = \frac{T_3' - T_2}{T_3 - T_2}$$

whence

$$T_3 - T_2 = (620 - 321.15)/0.80 = 374 \text{ K}$$

hence

$$T_3 = 374 + 321.15 = 695.15 \text{ K}$$

Example 6.2   A steam turbine plant works between the limits of 2 MN/m^2 and 573.15 K (300 °C) and 0.0035 MN/m^2. Compare the following four schemes;

(a) a superheated Rankine cycle;
(b) a reheat cycle, with the steam reheated to 573.15 K at the pressure where it becomes saturated;
(c) a regenerative cycle, with one bleed point (open heater) at the pressure where the steam becomes saturated;
(d) as in (c) but using a closed heater system.

(a)   Referring to Figure 6.31 steam tables show that

$$h_4 = 3025 \text{ kJ/kg}$$

$$s_4 = 6.768 \text{ kJ/kg K}$$

Since s_g at 0.0035 MN/m^2 = 8.521 kJ/kg K, steam is wet at point 5. Its dryness fraction is given by

$$x = \frac{s - s_f}{s_{fg}}$$

$$= \frac{6.768 - 0.391}{8.130}$$

$$= 0.785$$

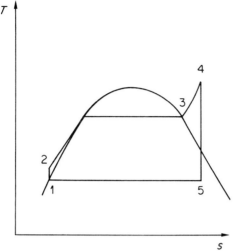

Figure 6.31

Therefore

$$h_5 = h_f + xh_{fg}$$

$$= 112 + 0.785 \times 2438$$

$$= 2023 \text{ kJ/kg}$$

hence

$$w_{45} = 3025 - 2023 = 1002 \text{ kJ/kg}$$

Now

$$w_{12} = h_1 - h_2$$

$$\simeq v_f(p_1 - p_2) = 0.0010(0.0035 - 2) \times 10^3 \text{ kJ/kg}$$

$$= -2 \text{ kJ/kg}$$

Therefore the net work output

$$w = w_{45} + w_{12} = 1000 \text{ kJ/kg}$$

Heat input

$$q_{42} = h_4 - h_2$$

But

$$h_2 = h_1 - w_{12} = 112 + 2 = 114 \text{ kJ/kg}$$

therefore

$$q_{42} = 3025 - 114 = 2911 \text{ kJ/kg}$$

whence

$$\eta = \frac{w}{q_{42}} = \frac{1000}{2911} = 0.344$$

Also

$$\text{s.s.c.} = \frac{3600}{w} = \frac{3600}{1000} = 3.6 \text{ kg/kW h}$$

(b)   Pressure at reheat point 5 can be found since

$$s_5 = s_{sat} = s_4 = 6.768 \text{ kJ/kg K}$$

therefore

$$p = 0.55 + \frac{6.768 - 6.790}{6.761 - 6.790}(0.6 - 0.55)$$

$$= 0.55 + \frac{22}{29} \times 0.5$$

$$= 0.588 \text{ MN/m}^2$$

hence

$$h_5 = 2753 + \frac{0.588 - 0.55}{0.60 - 0.55}(2757 - 2753)$$

$$= 2753 + \frac{0.038}{0.05} \times 4$$

$$= 2756 \text{ kJ/kg}$$

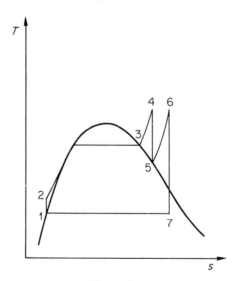

Figure 6.32

As 6 and 5 are on the same isobar

$$h_6 = 3065 + \frac{0.588 - 0.5}{0.60 - 0.5}(3062 - 3065)$$

$$= 3065 + \frac{0.088}{0.1}(-3)$$

$$= 3062.4 \text{ kJ/kg}$$

$$s_6 = 7.460 + 0.88 (7.373 - 7.460)$$

$$= 7.460 + 0.88 (- 0.087)$$

$$= 7.384 \text{ kJ/kg K}$$

Since s_g at $p = 0.0035 \text{ MN/m}^2$ is 8.521 kJ/kg K, steam is still wet on expansion to the exhaust pressure, and the dryness fraction is given by

$$x = \frac{7.384 - .391}{8.130} = 0.86$$

whence

$$h_7 = 112 + 0.86 (2438)$$

$$= 112 + 2095 = 2207 \text{ kJ/kg}$$

The net work output w is given by

$$w = w_{45} + w_{67} + w_{12}$$

$$= (3025 - 2765) + (3062.4 - 2207) - 2$$

$$= 1122.4$$

The heat input is

$$q = q_{42} + q_{65}$$

$$= 2911 + (h_6 - h_5)$$

$$= 2911 + (3062.4 - 2756)$$

$$= 3217.4$$

Therefore

$$\eta = \frac{1122.4}{3217.4} = 0.349$$

and

$$\text{s.s.c.} = \frac{3600}{w} = \frac{3600}{1122.4} = 3.2 \text{ kg/kW h}$$

(c)   As in (b)

$$w_{45} = 269 \text{ kJ/kg}$$

To determine the amount of steam bled off at 5, consider the heat balance for the open feed heater, where

$$1.h_{2'} = yh_5 + (1 - y) h_2$$

which gives

$$y = \frac{h_{2'} - h_2}{h_5 - h_2}$$

Now

$$h_{2'} = 656 + \frac{0.588 - 0.55}{0.60 - 0.55} \ (670 - 656)$$

$$= 656 + \frac{0.038}{0.05} \times 14$$

$$= 666.6 \ \text{kJ/kg}$$

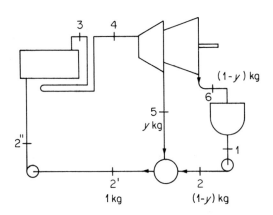

Figure 6.33

therefore

$$y = \frac{666.6 - 114}{2756 - 114}$$

$$= \frac{552.6}{2642} = 0.209 \ \text{kJ/kg}$$

whence

$$w_{56} = (1 - y) \ (h_5 - h_6)$$

$$= 0.791 \ (2756 - 2023)$$

$$= 580 \ \text{kJ/kg}$$

also

$$w_{2'2''} = v_{\text{f}} \ (p_{2'} - p_{2''})$$

$$= 0.0011 \ (0.588 - 2) \times 10^3$$

$$= -1.1 \times 1.412 = -1.55 \ \text{kJ/kg}$$

hence

$$w = w_{45} + w_{56} + w_{12} + w_{2'2''}$$

$$= 269 + 580 - 0.791 \times 2 - 1.55$$

$$= 845.87 \text{ kJ/kg}$$

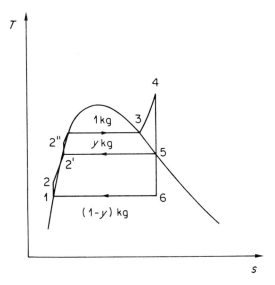

Figure 6.34

The heat input is

$$q_{42''} = 3025 - (666.6 + 1.55)$$

$$= 2356.8 \text{ kJ/kg}$$

whence

$$\eta = \frac{w}{q_{42''}} = \frac{845.87}{2356.8} = 0.3595$$

and

$$\text{s.s.c.} = \frac{3600}{w} = \frac{3600}{845.9} = 4.25 \text{ kg/kW h}$$

(d) As in (c)

$$w_{45} = 269 \text{ kJ/kg}$$

Heat balance for the heater as a closed system gives

$$1.h_{11} + yh_9 = 1.h_2 + yh_5$$

giving

$$y = \frac{h_{11} - h_2}{h_5 - h_9}$$

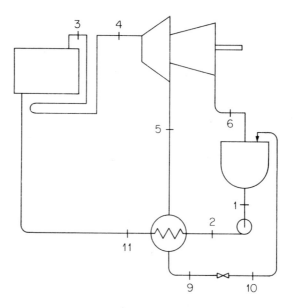

Figure 6.35

Now in finding the enthalpies in the feed line, it is usual to make the following assumptions;

(1) the feed pump term is negligible;
(2) the enthalpy of the compressed liquid is approximately equal to that of the saturated liquid at the same temperature;
(3) the states of the bled condensate before and after throttling is approximately equal to that of the saturated liquid at the lower pressure of the throttled liquid.

With the above assumptions, we have

$$h_2 = h_1$$
$$h_{11} = h_8$$
$$h_9 = h_{10} = h_1$$

whence

$$y = \frac{h_8 - h_1}{h_5 - h_1}$$

$$= \frac{666.6 - 112}{2756 - 112} = 0.209 \text{ kJ/kg}$$

Also

$$w_{56} = 580 \text{ kJ/kg}$$

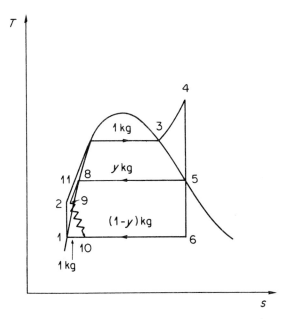

Figure 6.36

hence

$$w = w_{45} + w_{56} + w_{12}$$

$$= 269 + 580 - 2 = 847 \text{ kJ/kg}$$

Heat input $= q_{411} = 2358.4 \text{ kJ/kg}$
whence

$$\eta = \frac{w}{q_{411}} = \frac{847}{2358.4} = 0.360$$

and

$$\text{s.s.c.} = \frac{3600}{w} = \frac{3600}{847} = 4.25 \text{ kg/kW h}$$

It is therefore seen that there is no preference on thermodynamic grounds between open or closed feed heating. The choice is often made on capital outlay considerations: the closed system requires only one feed pump for the entire power plant, whereas the open system needs one feed pump for each feed heater in use.

Example 6.3 An air-standard Otto cycle inducts air at 288 K and 101.325 kN/m². The compression ratio of the cycle is 5. Heat addition amounts to 2600 kJ/kg of air inducted. Calculate the ideal cycle efficiency, the mean effective pressure and the peak pressure of the cycle.

By equation (6.22)

$$\eta = 1 - \frac{1}{5^{0.4}}$$

$$= 1 - \frac{1}{1.9} = 0.473$$

Since heat addition occurs at constant volume

$$\frac{p_3}{p_2} = \frac{T_3}{T_2}$$

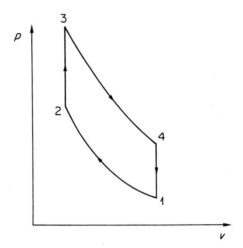

Figure 6.37

Assuming air behaves as perfect gas

$$q_{32} = c_v(T_3 - T_2)$$

therefore

$$\frac{T_3}{T_2} = 1 + \frac{q_{32}}{c_v T_2} = 1 + \frac{2600}{0.718 \, T_2}$$

But

$$\frac{T_2}{T_1} = \left(\frac{v_1}{v_2}\right)^{\gamma - 1} = 5^{0.4} = 1.904$$

therefore

$$T_2 = 288 \times 1.904 = 549 \text{ K}$$

whence

$$\frac{T_3}{T_2} = 1 + \frac{2600}{0.718 \times 549} = 7.60$$

hence

$$p_3 = p_2 \frac{T_3}{T_2} = p_1 \left(\frac{v_1}{v_2}\right)^{\gamma} \times 7.60$$

$$= 101.325 \times 5^{1.4} \times 7.60$$

$$= 7310 \text{ kN/m}^2 = 7.31 \text{ MN/m}^2$$

By equation (6.23)

$$p_m = \frac{(p_3 - p_1 r_c)}{(\gamma - 1)(r_c - 1)} = \frac{(7.31 - 0.101 \times 5^{1.4}) \times 0.473}{0.4 \times 4}$$

$$= 1.91 \text{ MN/m}^2$$

Example 6.4 A refrigerator having Freon–12 as working fluid works between pressure limits of 0.261 and 0.9607 MN/m². The refrigerant enters the compressor dry saturated. Find the refrigeration effect and the coefficient of performance if the refrigerant is subcooled to 293.15 K (20 °C) before throttling. Sketch the refrigeration cycle on both the p–h and T–s diagrams.

At $p = 0.261 \text{ MN/m}^2$, $T_{sat} = 268.15 \text{ K}$

$$h_g = h_1 = 185.38 \text{ kJ/kg}$$

$$s_g = s_1 = 0.6991 \text{ kJ/kg K}$$

At $p = 0.9607 \text{ MN/m}^2$,

$$s_g = 0.6825 \text{ kJ/kg K}$$

Hence, state 2 is superheated by

$$15 \times \frac{0.6991 - 0.6825}{0.7185 - 0.6825} = 15 \times \frac{0.0166}{0.0360}$$

$$= 6.91 \text{ K}$$

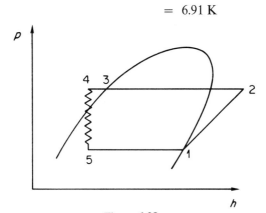

Figure 6.38

therefore

$$h_2 = 203.20 + \frac{6.91}{15}(214.76 - 203.20)$$

$$= 208.51 \text{ kJ/kg}$$

Now for a subcooled liquid

$$h_{\text{sub}} - h_{\text{sat}} = v_f(p_{\text{sub}} - p_{\text{sat}})$$

therefore

$$h_4 = h_3 + v_3(p_4 - p_3)$$

$$= 74.59 + \frac{1000}{1304}(0.9607 - 0.5673)$$

$$= 74.89 \text{ kJ/kg}$$

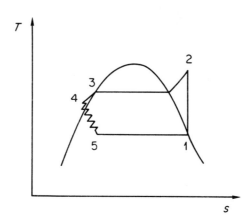

Figure 6.39

Since $h_5 = h_4$ for the throttling process

$$\text{R.E.} = h_1 - h_5$$

$$= 185.38 - 74.89 = 110.49 \text{ kJ/kg}$$

since

$$w = h_2 - h_1 = 208.51 - 185.38 = 23.13$$

$$\text{COP} = \frac{\text{R.E.}}{w} = \frac{110.49}{23.13} = 4.78$$

PROBLEMS

1. Solve Example 6.1 with steam instead of air as the working fluid.
2. Calculate the heat and work interactions, cycle efficiency, work ratio and steam consumption of a Carnot cycle, using steam between a saturation temperature of 503 K and a sink temperature of 300 K.
3. Resolve problem 2 with isentropic efficiencies of 0.80 for the compression and expansion processes.
4. (a) Calculate the cycle efficiency, work ratio, and steam consumption of the Rankine cycle working between the same conditions as the Carnot cycle in problem 1.
   (b) Estimate the actual cycle efficiency and steam consumption when the isentropic efficiencies of the expansion and compression processes are each 0.80.
5. Calculate the efficiency and specific work output of a simple gas turbine plant operating on the Joule or Brayton cycle. The maximum and minimum temperatures of the cycle are 1000 K and 290 K respectively; the pressure ratio is 6, and the isentropic efficiencies of the compressor and turbine stages are 85 and 90 percent respectively.

   What is the optimum pressure ratio for this gas turbine and what are the efficiency and work output under optimum operating conditions?
6. Show that for the Stirling cycle with all the processes occurring reversibly but where the heat rejected is not used for regenerative heating, the efficiency is given by

$$\eta = 1 - \frac{\left(\dfrac{T_1}{T_2} - 1\right) + (\gamma - 1)\ln r}{\left(\dfrac{T_1}{T_2} - 1\right) + (\gamma - 1)\dfrac{T_1}{T_2}\ln r}$$

   where r is the compression ratio and T_1/T_2 the maximum to minimum temperature ratio.

   Determine the efficiency of this cycle when using hydrogen ($R = 4.307$ kJ/kg K, $c_p = 14.50$ kJ/kg K) with a pressure and temperature prior to isothermal compression of 101.325 kN/m² and 300 K respectively, a maximum pressure of 2.55 MN/m² and heat supplied during the constant volume heating of 9300 kJ/kg of air ($c_p = 1.005$ kJ/kg K).

   If the heat rejected during the constant volume cooling can be utilized to provide the constant volume heating, what will be the cycle efficiency, other conditions remaining the same? Without altering the temperature ratio, could the efficiency be further improved by changes in the cycle?
7. Describe the sequence of operations in the conventional four-stroke Diesel engine and in the four-stroke petrol engine. Highlight the main differences.

   Upon what factor more than any other does the efficiency of an internal combustion engine depend?

8. A Carnot engine, having a perfect gas as working fluid, is driven backwards and is used for freezing water already at 273.15 K. If the engine is driven by a 500 W motor working at an efficiency of 65 percent, how long will it take to freeze 10 kg of water? Assume the engine rejects heat at a room temperature of 290 K and that there are no heat losses in the refrigerating system.

9. Calculate the refrigeration effect and coefficient of performance for a Freon-12 refrigerator, where the upper and lower temperatures of the cycle are 300 K and 258 K respectively. The refrigerant enters the compressor as a saturated vapour and is subcooled to 288 K before throttling.

10. The volumetric efficiency of a single stage reciprocating engine is defined as the ratio of the actual volume induced per cycle to the swept volume. Show that for a reversible compressor

$$\eta_{vol} = 1 - \frac{V_c}{V_s}\left\{\left(\frac{p_2}{p_1}\right)^{1/n} - 1\right\}$$

where V_c and V_s represent the clearance and swept volumes respectively.

Air is to be compressed in a single stage reciprocating compressor through a pressure ratio of 10:1. The induced air is at atmospheric pressure. The index of polytropic compression and expansion is 1.3. Calculate the power required for a free air delivery of 3 m³/min. (Free air delivery is defined as the rate of volume flow measured at inlet pressure and temperature.)

Show that in terms of easily measurable quantities, the volumetric efficiency and free air delivery is related by

$$V_f = \eta_{vol} N V_s$$

where V_f denotes the free air delivery, N the rotational speed of the compressor. Hence, find the rotational speed if the swept volume is 15 litres.

11. As the pressure ratio of a given reciprocating compressor is increased, the volumetric efficiency falls. Explain, with the aid of a p–v diagram, how this comes about. What form would a reciprocating compressor take if it was intended to produce a large pressure ratio without resorting to an impracticably long piston?

Show that in a two-stage compressor with complete intercooling after the first (or low pressure) stage, so that the temperature of the intermediately compressed gas returns back to that of the inlet value, the total work required is minimal if the work is divided equally between the stages. Assume the compression processes are reversibly polytropic.

12. Process efficiencies in general do not provide a good measure of the actual process, because this does not follow the path of the idealized process close enough to allow a fair comparison to be made. To overcome this shortcoming, the term *effectiveness* has come into wider usage. For an expansion or compression process, the following definitions are given

$$\varepsilon_{turb} = \frac{\text{actual work delivered}}{\text{maximum useful work deliverable (or decrease in availability)}}$$

$$\varepsilon_{comp} = \frac{\text{Maximum useful work deliverable (or increase in availability)}}{\text{actual work supplied}}$$

Using these definitions, solve example 6.1, assuming now that the effectiveness of each processes is 0.80. (As in the case of the definition for the process efficiencies, you may assume that these processes occur adiabatically.)

CHAPTER SEVEN

Properties of Real Substances

7.1 BEHAVIOUR OF REAL SUBSTANCES

In the discussion of ideal substances in chapter 5 it was assumed that substances exist permanently in one of the three states, solid, liquid or vapour. Property values at phase transitions are introduced via tabulated data. Obviously, such treatment of the behaviour of a substance is far from complete. It is the objective of the present chapter to continue the discussion from the point where it was left off in chapter 5. As an introduction to the behaviour of a real substance first consider the simple case of a kilogram of ice at a temperature of 260 K confined in a cylinder by a movable piston which exerts on the ice a pressure of one atmosphere, equivalent to a pressure of 101.325 kN/m². The temperature of all the ice may be raised uniformly by a slow addition of heat. The heat added is equal to the increase in enthalpy of the system, since the process is isobaric. When the temperature reaches 273.15 K, measurements would indicate that the enthalpy of the system has increased by about 25 kJ. Upon further heating, we find at first no further change in temperature, despite the continuous increase in enthalpy of the system. During this period, the substance exhibits a marked change in appearance; part of the ice has become liquid. The proportions of solid (ice) in the system can be changed at will by heating or cooling the system, though neither heating nor cooling will cause a change in temperature so long as ice and water are present together. When the enthalpy of the system has been increased by about 333 kJ over its value for the solid at 273.15 K, the entire mass will be in the liquid phase; further heating now results in a rise in temperature of the system.

The change in enthalpy between the solid and liquid phases at the same pressure and temperature has been called the *latent heat of fusion*—an unfortunate choice of words because the term heat should have been reserved for an interaction between systems by virtue of a temperature difference. Also, this change in enthalpy can be caused by means other than heat, such as a revolving paddlewheel in the liquid fraction or an electric current passing through the substance. We shall nevertheless retain this commonly accepted term.

Upon further heating, the liquid continues to rise in temperature until 373.15 K. At this temperature, its enthalpy is some 418 kJ higher than that of the liquid at 273.15 K. Once again, we reach a point where the enthalpy may be changed without change in temperature; and once more we notice a change in the appearance of the substance as the enthalpy is increased—some of it becomes vapour. The proportions of vapour in the system can be changed at will by heating and cooling and neither heating nor cooling will cause a change in its temperature so long as both phases coexist. When the enthalpy has been increased by some 2260 kJ over its value for the liquid at 373.15 K, the entire mass will will be in the vapour phase and further heating will result in a rise in temperature.

The change in enthalpy between the liquid and vapour phases at the same pressure and temperature is called the *latent heat of vaporization*. The liquid at 1 atm and 373.15 K, from which the vapour is evaporated and with which vapour can exist in all proportions in equilibrium, is an example of saturated liquid. The vapour which is formed from the liquid and with which the liquid can exist in equilibrium in all proportions is an example of saturated vapour. Further heating will cause a rise in the temperature of the vapour which is then known as superheated vapour.

If the pressure exerted by the piston is varied within certain limits, the phenomenon remains unchanged qualitatively; only the quantitative values of temperature and enthalpies for the phase changes become different. When such an experiment has been conducted a sufficient number of times for various pressures, the pressure–temperature dependence for the substance may be constructed in the form shown in Figure 7.1.

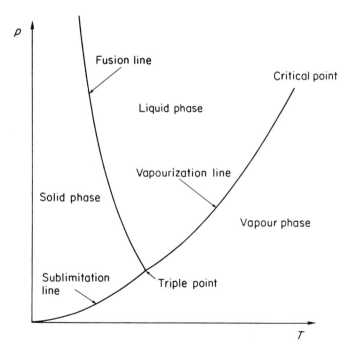

Figure 7.1

From this figure, it is seen that below a certain pressure the solid phase may exist in equilibrium with the vapour phase, i.e. it may be converted directly to the vapour phase without passing through the liquid stage (a process called sublimation). Above this pressure, the solid can exist in equilibrium only with the liquid phase. The point at which all three phases may exist in equilibrium is called the triple point. The vaporization line ends at the critical point beyond which there is no latent heat of vaporization and no other characteristic change which normally marks a change in phase.

Because of the latent heats, the phase boundaries which are lines on the temperature–pressure diagram are bands on the enthalpy–pressure diagram. Also, owing to the change in volume during phase transitions, the phase boundaries are areas on the pressure–volume diagram too. We have already come across the p–h diagram in conjunction with the discussion on refrigeration cycles (see Figure 6.27). A typical p–v diagram for a fluid near the two-phase region is shown in Figure 7.2.

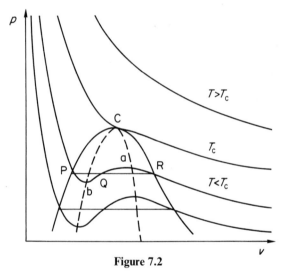

Figure 7.2

7.2 DENSE GASES: VAN DER WAALS' EQUATION

Reference to Figure 7.2 shows that the semiperfect and perfect gas approximations apply only when v is large, i.e. has a low density, or when T is high ($\approx 3T_c$). Under such conditions, the average distance between molecules is large, so that the energy associated with attractive forces between molecules can be neglected.

Van der Waals suggested a modification of the ideal gas equation to account for the volume occupied by the molecules and the intermolecular forces of attraction. This has the form

$$\left(p + \frac{a}{\tilde{v}^2}\right)(\tilde{v} - b) = \hat{R}T \tag{7.1}$$

where $\tilde{v}$ is the molar volume and a and b are constants for particular substances.

In spite of its simplicity, the van der Waals' equation comprehends both the gaseous and liquid phases and brings out, in a most remarkable way, all the phenomena pertaining to the continuity of these two phases. Isotherms obtained from this equation appear as shown in Figure 7.2. For $T > T_c$, these are not unlike those given by the ideal gas equation and correspond therefore to the gaseous state of the substance. Below T_c, each isotherm passes through a maximum and a minimum in an intermediate region, which can be taken to correspond to the two-phase transitional region between the gaseous and liquid phases. The locus of these extreme points can be obtained by setting $(\partial p/\partial v)_T = 0$. It is given by the equation

$$p = \frac{a}{\tilde{v}^2} \left\{ \frac{2(\tilde{v} - b)}{\tilde{v}} - 1 \right\} \qquad (7.2)$$

This is shown by the dotted curve in Figure 7.2. By the following reasoning, the portion of the isotherms within this dotted curve can be shown to have no physical significance. Imagine that the fluid in a state corresponding to this part of the curve is contained in a diathermal cylinder fitted with a vertical piston weighted in equilibrium with the fluid pressure. If we take away a little weight form the piston, there will no longer be equilibrium and it will begin to rise. However, as it rises its volume increases and with it also its pressure. The resultant force on the piston further increases and the gas continues to expand until it reaches the maximum of the isotherm. *Vice versa*, if we add a little weight onto the piston initially, the gas will collapse to the state corresponding to the minimum of the isotherm. This portion, therefore, represents an unstable state of the fluid. Leaving out this unrealistic portion, isotherms for $T < T_c$ can be broken up into two branches: the one of large molal volume corresponding to the gaseous state, referred to earlier; and that of small molal volume representing the liquid state. Our experiment in Section 7.1 has indicated that in the two phase region, the isotherms should be horizontal. Nevertheless, the portions of the isotherm lying between Pb and aR have physical significance. The former represents states of the liquid for which the temperature exceeds the saturation temperature corresponding to that pressure. These are known as superheated states of the liquid. Similarly, the latter represents states of the vapour for which the temperature is less than the saturation temperature corresponding to that pressure. These are said to be supersaturated or subcooled states of the vapour. Both are metastable states and will return immediately to the thermodynamically stable states on line PR when the system is slightly disturbed.

Furthermore, we can use the second law to deduce the saturation curve on the p–v diagram of a van der Waals fluid, in the following way. A change of state can always be brought about at constant pressure and temperature from one phase to another for any real fluid—the process for this change corresponds to PQR shown in Figure 7.2. For a van der Waals fluid, there exists a path at constant temperature RaQbP by which the fluid can be returned to its original state. As this cycle can be executed reversibly at only one level of temperature, it is necessary according to the second law that the net work done by the fluid should be zero. Thus,

$$\oint p\,d\tilde{v} = 0$$

It follows therefore that areas PQbP and QaRQ must be equal. This condition allows us to identify the saturated states on the van der Waal isotherms. The state corresponding to the point C at which no distinction exists between the gaseous and liquid states is called the critical point. The isotherm through this point, T_c, exhibits a point of inflexion, at which both $(\partial p/\partial \tilde{v})_T$ and $(\partial^2 p/\partial \tilde{v}^2)_T$ are zero. These conditions together specify the position of this point as

$$\tilde{v}_c = 3b$$

$$p_c = \frac{a}{27b^2} \tag{7.3}$$

and

$$T_c = \frac{8a}{27Rb}$$

We can further eliminate the constants a and b among these relations to obtain

$$z_c = \frac{p_c\,\tilde{v}_c}{\tilde{R}\,T_c} = \frac{p_c\,v_c}{R\,T_c} = \frac{3}{8} \tag{7.4}$$

Thus, the value of z at the critical state is independent of the magnitude of the van der Waals' constants. The parameter z is known as the compressibility factor. For an ideal gas, it has a value of unity.

Using the set of critical constants found above, we can rewrite van der Waals' equation in a normalized form which contains no arbitrary constants. This is

$$\left(p^* + \frac{3}{v^{*2}}\right)(3v^* - 1) = 8T^* \tag{7.5}$$

Theoretically, this equation should be applicable to all substances. The parameters p^*, v^* and T^* are called respectively the reduced pressure, volume and temperature. States of two substances having the same values of p^*, v^* and T^* are called corresponding states. The above equation expresses what has come to be known as the *law of corresponding states*: If two of the reduced parameters are the same for different substances, the third is also the same. The mathematical reason behind the validity of this law lies in the fact that the van der Waals equation contains just as many constants as variables; namely, three each. The derivation of this from the van der Waals equation is a fortuitous coincidence, since it turns out to have a wider validity than the equation from which it is derived.

7.3 LAW OF CORRESPONDING STATES

The reduced van der Waals' equation is of great value when examining qualitatively the properties of fluids. Quantitative predictions are not sufficient for
accurate work. Realizing this, van der Waals suggested an empirical way of
correlating the p–v–T data based on the law of corresponding states deduced
from his equation of state. This can be expressed mathematically as

$$f\,(p^*,\; v^*,\; T^*)\; =\; 0 \tag{7.6}$$

One way of presenting this three-dimensional relation in a two-dimensional
form is to plot p^* and T^* as coordinates, with v^* as the parameter. This form of
plotting the results is equivalent to writing equation (7.6) in the form

$$v^*\; =\; f\,(p^*,\; T^*) \tag{7.7}$$

Thus for each value of v^* the p^* *versus* T^* plot should yield a single curve for all
substances. This curve is known as a reduced isometric. The entire plot consists,
therefore, of a family of reduced isometrics. Experimental data for a number of
hydrocarbons at pressures and temperatures that are high relative to the critical
pressure and temperature give a fair degree of accuracy ($\approx$ 10 percent).
    This law breaks down near the critical region since the parameter z_c for real
substances varies from 0.18 to 0.30. It also breaks down in the ideal gas region,
where $z = 1$. This can be seen from the following consideration.
Since

$$\frac{p^*\,v^*}{T^*}\; =\; \frac{pv}{RT}\; \times\; \frac{RT_c}{p_c\,v_c}$$

In the ideal gas region, this becomes

$$\frac{p^*\,v^*}{T^*}\; =\; \frac{RT_c}{p_c\,v_c} \tag{7.8}$$

As the LHS is a function of p^*, v^* and T^* and the RHS a function of p_c, v_c and
T_c, in order that this relation be valid for all substances, each must be constant.
The right hand side parameter, however, equals $1/z_c$ which varies since the
denominator varies.
    Because of these factors, the above method of presenting the law of corre-
sponding states is not satisfactory.
    An alternative form of correlating the data is to introduce the factor z into
the plot. This can be achieved by formally multiplying equation (7.7) on both
sides by the factor p^*/T^*, which then gives

$$\frac{z}{z_c}\; =\; g\,(p^*,\; T^*) \tag{7.9}$$

This relation allows the continuous variation of z to be displayed and is thus better than the previous form in that now the ideal gas region can be made valid, even if z_c has to be taken as a constant. A chart displaying the variations among z, p^* and T^* is known as the *generalized compressibility chart*. In actual fact, due to the variation in z_c, high accuracy in the critical region cannot be attained with a plot with z_c taken to be constant. To account for this, four charts with different values for z_c have been reported. These are;

    I: $0.24 < z_c < 0.26$, average of 0.25, for substances such as acetone, ammonia, esters, alcohols;

    II: $0.26 < z_c < 0.28$, average of 0.27, mostly for hydrocarbons;

    III: $0.28 < z_c < 0.30$, average of 0.29, for oxygen, nitrogen, carbon monoxide, methane, ethane, argon, hydrogen sulphide;

    IV: $z_c = 0.23$, for water.

More than 62 percent of the 82 materials used to develop the above correlations fall in Group II. This particular chart can therefore be used to represent with reasonable accuracy a wide range of real gases. It has often been referred to as *the* generalized compressibility chart. Students should however realize that for accurate calculations, a separate chart is necessary for each group.

The liquid region is usually not included in compressibility charts because liquid compressibilities do not 'generalize' well as functions of p^* and T^* only.

The generalized chart method suffers from one numerical drawback in that if v is one of the given parameters of state, successive approximation procedures have to be used in solving the problem. In addition, the determination of the other reduced parameters involves the use of the expression

$$\frac{z}{z_c} = \frac{p^* v^*}{T^*} \tag{7.10}$$

which introduces further inconveniences. These are (1) the value of z_c can be subjected to large errors (of the order of ± 25 percent) and (2) the value v_c for the substance is not always known and if known may not be consistent with the generalized chart. To overcome these difficulties, we define a *pseudo-reduced volume* by the relation

$$v^{*\prime} = \frac{p_c v}{R T_c} \tag{7.11}$$

When used in plotting the generalized charts, we are able to specify the third (unknown) variable knowing any two properties with relative ease.

7.4 OTHER EQUATIONS OF STATE

An equation of state is any expression relating the three primitive properties p, v and T. We have come across two such equations—the ideal gas equation and van der Waals' equation. Though it is impossible to deduce these relations from the laws of thermodynamics, they are undoubtedly useful in helping us to understand the behaviour of the substances constituting our simple systems which are our main concern. Furthermore, they are necessary for extending the often inadequate p–v–T data into other ranges where values for derived properties are to be calculated. Over 150 such equations have been proposed. A few of these that are more frequently encountered are listed below. Needless to say, all these approach the ideal gas equation as $p \to 0$. It is also worth noting that many of these revert to van der Waals' equation with slight restrictions.

Berthelot's equation

$$\left(p + \frac{a}{Tv^2}\right)(v - b) = RT \tag{7.12}$$

This equation is much used by physical chemists. It has the same form as van der Waals' equation. The only difference is that now the constant a is taken to vary inversely as temperature. The reduced equation obtained from this does not satisfy the critical state.

Dieterici equation

$$p(v - b) = RT\,e^{-a/RTv} \tag{7.13}$$

The Dieterici equation was proposed to meet the objection that z_c of van der Waals' equation is too high for any real substance. The value of z_c for this equation is about 0.27, which is the average of values for real substances. It yields isotherms on a p–v diagram which resemble those of van der Waals in most respects. It differs, however, in that the pressure is never less than zero. This is not necessarily an advantage, in that real substances actually exhibit states for which the pressure is negative.

In practice, the Dieterici equation has proved to be superior to the van der Waals equation; particularly in the neighbourhood of the critical point, where it more faithfully follows the characteristics of real substances. It reduces to van der Waals' equation for small a and b.

Saha–Bose equation

$$p = -\frac{RT}{2b}e^{-a/RTv}\ln\left(\frac{v - 2b}{v}\right) \tag{7.14}$$

This contains the van der Waals equation as a special case.

The above equations have only two constants (not including R), and satisfy the law of corresponding states. They cannot, therefore, be accurately representative and each is valid over a limited range only.

Equations listed below contain more arbitrary constants than van der Waals' equation and are therefore of more general validity.

142

Clausius equation

$$\left(p + \frac{a}{T(v+c)^2}\right)(v-b) = RT \qquad (7.15)$$

(3 constants).

Beattie–Bridgeman equation

$$p = \frac{RT}{v^2}\left(1 - \frac{c}{vT^3}\right)\left(v + B_0 - \frac{bB_0}{v}\right) - \frac{A_0}{v^2}\left(1 - \frac{a}{v}\right) \qquad (7.16)$$

(5 constants). Alternatively, this can be written as

$$p = \frac{RT}{v} + \frac{\alpha}{v^2} + \frac{\beta}{v^3} + \frac{\gamma}{v^4}$$

where

$$\alpha = B_0 RT - A_0 - \frac{cR}{T^2}$$

$$\beta = -B_0 bRT + A_0 a - \frac{B_0 cR}{T^2}$$

$$\gamma = \frac{B_0 bcR}{T}$$

The Beattie–Bridgeman equation is quite accurate for densities less than about 0.8 times the critical density.

All the equations listed above reduce to the ideal gas equation for large v and T. All those equations which do not involve temperature effects in the correction factors reduce also to the exact form of van der Waals' equation on certain simplifying assumptions. Others incorporating the temperature effects can be viewed as empirical modifications of van der Waals' equation.

Virial forms of the equation of state

In these forms, the actual number of arbitrary constants can be chosen so that agreement with experimental data is as closed as desired. The virial equation of state can be written in one of the two forms given below:

$$z = \frac{pv}{RT} = 1 + \frac{B(T)}{v} + \frac{C(T)}{v^2} + \frac{D(T)}{v^3} + \dots \qquad (7.17a)$$

or

$$z = \frac{pv}{RT} = 1 + B'(T)p + C'(T)p^2 + D'(T)p^3 + \dots \qquad (7.17b)$$

The coefficients $B(T)$, $B'(T)$, etc., of the virial equations are temperature dependent functions. They are called the virial coefficients. It can be shown, by methods of statistical mechanics, that the second virial coefficients B or B', have their origins in binary molecular collisions, the third in tertiary collisions, etc. Where the colliding molecules are few, it is possible to predict these coefficients from statistical considerations together with empirical relations for the molecular force field. Using up to the third virial coefficient, it is possible to achieve accurate p–v–T data representation at densities below about 0.7 times the critical density.

The coefficients in the two virial forms of the equation of state bear simple relationships to each other, e.g.

$$B' = \frac{B}{RT} \text{ and } C' = \frac{C - B^2}{(RT)^2} \tag{7.18}$$

The second form (equation 7.17b) has the advantage that the independent variables are the easily measurable quantities of pressure and temperature. This advantage is often offset, however, by the greater ease with which the first form can be made to fit the characteristics of real substances. The fact that the van der Waals' as well as all other equations of state given above are of the form

$$p = p(v, T)$$

is evidence of the relative facility with which this form can be used.

7.5 DEVELOPING THERMODYNAMIC PROPERTIES FROM EXPERIMENTAL DATA

Up to now the behaviour of the primitive properties of a real substance have been considered. Derived properties are in fact more important as regards thermodynamic calculations. For common substances, such as H_2O, NH_3 and Freon 12, these properties are commonly available either graphically or in table form. However, occasion may arise when the development of derived properties for a material is necessary. The following discussion gives a brief summary of the various methods that can be used for developing these properties for a one-component simple system.

Broadly speaking, such methods fall into one of two categories. The first is used for substances for which experimental p–v–T data are available, even though these frequently only exist over fairly narrow ranges of temperature and pressure. In order to develop properties over a sufficiently adequate range, these data have to be extended by equations of state that are found to be suitable. The second approach is generally used for substances with no experimental p–v–T data. For these substances, it is usual practice to resort to approximate information provided by the generalized compressibility or related charts based on the law of corresponding states.

In addition to p–v–T data, auxiliary experimental data are also required for evaluating the functions of integrations at the reference temperature. These are of the two following types.

(i) Heat capacity data. These are usually available over some specified temperature range, for either constant pressure or constant volume. They are either obtained from direct calorimetric measurements or from analysis of spectroscopic data.

(ii) Joule–Kelvin (Thomson) data. These exist for a few materials only, as experimental measurements are relatively difficult to make.

Derived properties of interests are s, h, u, a and g. Since a and g can be computed from known values of $p, v, T,$ and s together with either u or h, and since u and h are themselves related by the expression $h = u + pv$, it suffices, therefore, to present either u or h together with s. The choice between u or h is merely a matter of convenience, depending on the existence of auxiliary data and the explicit form of the equation of state. The usual practice is to tabulate h, which is rather more commonly encountered in engineering analysis. We shall describe below, for the sake of completeness, methods for developing all three properties, s, h and u.

Entropy

From the fourth Maxwell equation

$$\left(\frac{\partial s}{\partial p}\right)_T = -\left(\frac{\partial v}{\partial T}\right)_p \tag{3.41}$$

we can integrate along an isotherm to give

$$s(p, T) - s(p_1, T) = - \int_{p_1}^{p} \left(\frac{\partial v}{\partial T}\right)_p dp \bigg|_T \tag{7.19}$$

Suppose heat capacity data are available at p_1, we have

$$\left(\frac{\partial s}{\partial T}\right)_{p_1} = \frac{c_{p_1}}{T} \tag{7.20}$$

which yields on integration

$$s(p_1, T) = \int_{T_1}^{T} \frac{c_{p_1}}{T} dT + s(p_1, T_1) \tag{7.21}$$

hence

$$s(p, T) = - \int_{p_1}^{p} \left(\frac{\partial v}{\partial T}\right)_p dp \bigg|_T + \int_{T_1}^{T} \frac{c_{p_1}}{T} dT + s(p_1, T_1) \tag{7.22}$$

If Joule–Kelvin data are used instead, then from the definition of the Joule–Kelvin coefficient, we have

$$\mu(p, T) = \frac{T^2}{c_p}\left[\frac{\partial}{\partial T}\left(\frac{v}{T}\right)\right]_p \tag{3.57}$$

$$= -\frac{1}{c_p}\frac{\partial(v\tau)}{\partial\tau}\bigg|_p \quad \text{where} \quad \tau = \frac{1}{T}$$

whence

$$C_{p_1} = -\frac{\dfrac{\partial(v\tau)}{\partial\tau}\bigg|_{p_1}}{\mu(p_1, T)} \tag{7.23}$$

and the expression for s becomes

$$s(p, T)\int_{p_1}^{p}\left(\frac{\partial v}{\partial T}\right)_p \mathrm{d}p\bigg|_T - \int_{T_1}^{T}\frac{1}{\mu(p_1, T)}\frac{\partial(v\tau)}{\partial\tau}\bigg|_{p_1}\frac{\mathrm{d}T}{T}$$

$$+ s(p_1, T_1) \tag{7.24}$$

The above expressions are convenient when the equation of state is explicit in v, i.e. of the form $v = v(p, T)$. For equations of state in the form $p = p(v, T)$, it is more convenient for us to develop s in a different manner, using the third Maxwell equation

$$\left(\frac{\partial s}{\partial v}\right)_T = \left(\frac{\partial p}{\partial T}\right)_v \tag{3.38}$$

as the starting point. Following the same procedure as used above, this gives

$$s(v, T) = \int_{v_1}^{v}\left(\frac{\partial p}{\partial T}\right)_v \mathrm{d}v\bigg|_T + \int_{T_1}^{T}\frac{c_{v_1}}{T}\mathrm{d}T + s(v_1, T_1) \tag{7.25}$$

This form is particularly convenient if we are to tabulate s and u.

Usually the equation of state is not simple enough to permit the first integral on the right to be evaluated analytically. Its existence implies that curves of constant pressure or volume can be plotted on a T–v or T–p diagram. From the slopes of such curves the necessary derivatives may be read off and the integration evaluated numerically.

In order to develop entropy data from the generalized compressibility chart, we again make use of equation (7.19). Since now

$$pv = zRT \tag{7.26}$$

we have

$$\left(\frac{\partial v}{\partial T}\right)_p = \frac{zR}{p} + \frac{RT}{p}\left(\frac{\partial z}{\partial T}\right)_p = \frac{R}{p}\left\{z + T\left(\frac{\partial z}{\partial T}\right)_p\right\} \tag{7.27}$$

$$= R\left\{z + T^*\left(\frac{\partial z}{\partial T^*}\right)_{p^*}\right\}\frac{1}{p_c p^*}$$

whence

$$s(p, T) - s(p_1, T) = -R\int_{p_1^*}^{p^*}\left\{z + T^*\left(\frac{\partial z}{\partial T^*}\right)_{p^*}\right\}\frac{dp^*}{p^*} \tag{7.28}$$

In practice, the difference of entropies between the perfect and real gases at the same temperature and pressure is used. For this purpose, we can let $p_1 \to 0$ in the above expression, and assume the state corresponding to zero pressure behaves as a perfect gas. Now from equation (5.15), for a perfect gas

$$s'(p, T) - s'(p_1, T) = R\ln\frac{p_1}{p} = -R\int_{p_1^*}^{p^*}\frac{dp}{p} \tag{7.29}$$

Thus on subtracting equation (7.28) from equation (7.29)

$$s'(p, T) - s(p, T) = R\int_0^{p^*}\left\{z - 1 + T^*\left(\frac{\partial z}{\partial T^*}\right)_{p^*}\right\}\frac{dp^*}{p^*} \tag{7.30}$$

since $s'(p_1, T) = s(p_1, T)$ as $p_1 \to 0$.

The term on the right of equation (7.30) is called the entropy departure, usually denoted by $\Delta s_{p, T}$. A graph for this quantity is available. This practically relieves us of all tedium involved in the evaluation of the integral in equation (7.30). All that remain for us to do when the entropy difference between two arbitrary states of a real gas is required is to apply the so-called three-step process. This is illustrated by the following example:

$$s(p_2, T_2) - s(p_1, T_1) = -\{s'(p_2, T_2) - s(p_2, T_2)\}$$

$$+ \{s'(p_2, T_2) - s'(p_1, T_1)\}$$

$$+ \{s'(p_1, T_1) - s(p_1, T_1)\}$$

$$= -\Delta s_{p_2, T_2} + \Delta s' + \Delta s_{p_1, T_1} \tag{7.31}$$

Alternatively, when proceeding from equation (7.25), we can evaluate the entropy difference as a function of volume and temperature. The compressibility chart is not applicable for the present purpose since this is plotted in the z–p^* coordinates. However, we can resort to the reduced isometric chart mentioned earlier. Observing that the reduced isometrics are straight lines of the form

$$p^* = (KT^* + L)_{v^*} \tag{7.32}$$

where K and L are functions of v^* only
then

$$(s_{v_2^*} - s_{v_1^*})_{T^*} = \frac{p_c v_c}{T_c} \int_{v_1^*}^{v_2^*} \left(\frac{\partial p^*}{\partial T^*}\right)_{v^*} dv^*$$

$$= \frac{p_c v_c}{T_c} \int_{v_1^*}^{v_2^*} K\, dv^* \tag{7.33}$$

This equation indicates that the entropy change of an isothermal process between two reduced states having volumes v_2^* and v_1^* is always the same regardless of the temperature level. The assumptions whereby this result is arrived at are (i) the substance satisfies the law of corresponding states and (ii) the reduced isometrics are linear.

Since the law of corresponding states does not hold at low pressures, it would be best for the low pressure region to be excluded from the plot for reduced isometrics. However, it can be readily shown that linear transformations on either the p^* or T^* coordinates have no effect on the expression for entropy difference as given by equation (7.33).

Enthalpy

Starting from the defining relation for h

$$dh = Tds + vdp \tag{3.33}$$

we obtain

$$\left(\frac{\partial h}{\partial p}\right)_T = v - T\left(\frac{\partial v}{\partial T}\right)_p = \frac{\partial (v\tau)}{\partial \tau}\bigg|_p \tag{7.34}$$

which on integration yields

$$h = \int_{p_1}^{p_2} \left. \frac{\partial (v\tau)}{\partial \tau} \right|_p dp \bigg|_T + F(T) \qquad (7.35)$$

With supplementary heat capacity data

$$F(T) = \int_{T_1}^{T} c_{p_1} dT + h(p_1, T_1) \qquad (7.36)$$

Alternatively, if auxiliary Joule–Kelvin data is used, then

$$F(T) = - \int_{T_1}^{T} \frac{\left. \frac{\partial (v\tau)}{\partial \tau} \right|_{p_1}}{\mu(p_1, T)} dT + h(p_1, T_1) \qquad (7.37)$$

If p–v–T data are not available, using the generalized compressibility chart, equation (7.34) can be written as

$$\left(\frac{\partial h}{\partial p} \right)_T = \frac{zRT}{p} - \frac{R}{p} \left\{ zT + T^2 \left(\frac{\partial z}{\partial T} \right)_p \right\} = \frac{RT^2}{p} \left(\frac{\partial z}{\partial T} \right)_p \qquad (7.38)$$

whence

$$h(p, T) - h(p_1, T) = RT_c T^{*2} \int_{p_1^*}^{p^*} \left(\frac{\partial Z}{\partial T^*} \right)_p \frac{dp^*}{p^*} \qquad (7.39)$$

At this stage, we can use similar arguments to those used earlier for entropy, to introduce the enthalpy departure. However, our task is much simpler this time if we realize that .pressure has no effect on the enthalpy of a perfect gas. Thus, let $p_1 \to 0$ and at the same time write

$$h(p_1, T) = h(0, T) = h'(p, T) \qquad (7.40)$$

Thus, equation (7.39) becomes

$$\frac{\Delta h_{p,T}}{T_c} = - RT^{*2} \int_0^{p^*} \left(\frac{\partial Z}{\partial T^*} \right)_{p^*} \frac{dp^*}{p^*} \qquad (7.41)$$

A graph for the quantity $\Delta h_{p,T}/T_c$ is available. To compute the enthalpy difference between two arbitrary states of a real gas, the three-step process discussed for finding the entropy difference can be applied.

Internal energy

Starting from the defining equation

$$du = T ds - p dv \tag{3.28}$$

it follows that

$$\left(\frac{\partial u}{\partial v}\right)_T = T\left(\frac{\partial s}{\partial v}\right)_T - p = T\left(\frac{\partial p}{\partial T}\right)_v - p = \left(\frac{\partial(p\tau)}{\partial\tau}\right)_v \tag{7.42}$$

hence

$$u(v, T) = -\int_{v_1}^{v}\left(\frac{\partial(p\tau)}{\partial\tau}\right)_v dv + F'(T) \tag{7.43}$$

Using auxiliary heat capacity data

$$F'(T) = \int_{T_1}^{T} c_{v_1} dT + u(v_1, T_1) \tag{7.44}$$

In the absence of p–v–T or specific heat at constant volume data, we can employ the reduced isometric chart for the development of values for internal energy. (The generalized compressibility chart is not used for the same reason as discussed in entropy data development earlier.) Thus equation (7.42) gives

$$\left(\frac{\partial u}{\partial v}\right)_T = T^* p_c \left(\frac{\partial p^*}{\partial T^*}\right)_{v^*} - p_c p^*$$

$$= p_c \left\{ T^* \left(\frac{\partial p^*}{\partial T^*}\right)_{v^*} - p^* \right\}$$

$$= p_c (T^* K - p^*) = -p_c L \tag{7.45}$$

where

$$L = T^* K - p^*.$$

Hence

$$u(v, T) - u(v_1, T) = -p_c v_c \int_{v_1^*}^{v^*} L dv^* = p_c v_c \int_{v^*}^{v_1^*} L dv^* \tag{7.46}$$

The above expression can be easily extended into the low pressure region where $v_1^* \to \infty$ because the asymptotic nature of the plot L versus v^*. Notice that as in the case for entropy changes, the internal energy change is not a function of reduced temperature either. Thus, no difficulty will be experienced in the low T^* range.

Specific heats

So far in our discussion, it has been seen that specific heat data play a predominant role in evaluating all derived properties. At first look, it may appear that a large amount of such data is needed to facilitate the arguments developed above since the reference state 1 is rather arbitrary, and is fixed by the availability of specific heat data at that pressure. In practice, it turns out that all that is needed is the specific heat at zero pressure, c_{po}. When this is known, then the specific heat at any pressure can be obtained by integrating the relation

$$\left(\frac{\partial c_p}{\partial p}\right)_T = - T\left(\frac{\partial^2 v}{\partial T^2}\right)_p \tag{7.47}$$

to give

$$c_p - c_{po} = - T \int_0^p \left(\frac{\partial^2 v}{\partial T^2}\right)_p \, dp \tag{7.48}$$

From the value of c_p, a value for c_v can be obtained using the relation for the difference in specific heats, i.e.

$$c_p - c_v = T \left(\frac{\partial v}{\partial T}\right)_p \left(\frac{\partial p}{\partial T}\right)_v \tag{7.49}$$

The two-phase region

The development of our previous discussion depends on the existence of the various partial derivatives. In a region consisting of a single phase only, these derivatives are well defined and the above techniques are applicable. However, such is not the case in the two-phase region which possesses a distinct discontinuity in the slopes of most properties. It is therefore necessary for us to seek a different approach to the problem. We have earlier seen in Section 3.9 that the Clapeyron equation relates certain properties during a change of phase. It is given by

$$\frac{dp}{dT} = \frac{h_{fg}}{T(v_g - v_f)} \tag{3.55}$$

This can be used at least for determining the latent enthalpy h_{fg} if the other quantities can be measured. On the left hand side of this equation, we have the slope of the pressure–temperature curve for the saturated substance. This can be obtained from the vapour-pressure curve, which is of the form

$$\ln p = A + \frac{B}{T} + C \ln T + DT \tag{7.50}$$

This equation has been found to give adequate representation of vapour-pressure data in a piecewise manner. Thus, by differentiating this equation, the LHS of the Clapeyron equation can be obtained. Now on the RHS, T is the saturation temperature, which is a known quantity. From the vapour-pressure equation, we can then compute the corresponding saturation pressure. These two quantities, when substituted into the equation of state of the vapour phase, allow the quantity v_g to be determined. Now v_f can always be measured—it is the reciprocal of the density of the saturated fluid. Thus h_{fg} can be found by calculation.

From this, the entropy change for the phase transition can be found by the relation

$$s_{fg} = \frac{h_{fg}}{T} \qquad (7.51)$$

and the internal energy, if required, can be obtained from the relation

$$h_{fg} = u_{fg} + p_{sat} (v_g - v_f) \qquad (7.52)$$

From the foregoing discussion, it is seen that the following data for any substance are necessary for developing tables of its derived properties;

(1) vapour-pressure data over a wide range;
(2) p–v–T data in the vapour region;
(3) the density of the saturated liquid and the critical pressure and temperature;
(4) zero pressure specific heat for the vapour or Joule–Kelvin data.

EXAMPLES

Example 7.1 Calculate the work and heat interactions for 1 kg of ethane when compressed reversibly and isothermally from 1 atm (101.325 kN/m²) to 68 atm at a temperature of 317 K in a steady flow process.

For ethane, $T_c = 305.48$ K, $p_c = 48.20$ atm, $z_c = 0.285$, therefore

$$p_{r_1} = \frac{1}{48.2} = 0.0208$$

$$T_{r_1} = T_{r_2} = \frac{317}{305.48} = 1.037$$

$$p_{r_2} = \frac{68}{48.2} = 1.412$$

From the generalized enthalpy departure chart

$$\frac{\tilde{h}_1^* - \tilde{h}_1}{T_c} < 0.01$$

152

$$\frac{\tilde{h}_2^* - \tilde{h}_2}{T_c} = 31.4$$

where h^* denotes enthalpy as $p \to 0$. For the present case, $\tilde{h}_1^* = \tilde{h}_2^*$ since as $p \to 0$, the enthalpy is a function of T only.

Setting $\tilde{h}_1 \simeq \tilde{h}_1^*$

$$\tilde{h}_2^* - \tilde{h}_2 \simeq \tilde{h}_1 - \tilde{h}_2 = 31.4 \times 305.48 = 9060 \text{ J/mol}$$

Since molecular weight of ethane (C_2H_6) is 30

$$h_1 - h_2 = \frac{9060}{30} = 302 \text{ kJ/kg}$$

Now the work interaction during isothermal compression is

$$w = RT \ln \frac{f_1}{f_2}$$

From the generalized fugacity chart

$$\frac{f_1}{p_1} = 1.0$$

$$\frac{f_2}{p_2} = 0.58$$

Therefore

$$w = \frac{8.3143}{30} \times 305.48 \ln \frac{1}{68 \times 0.58} \text{ kJ/kg}$$

$$= -311 \text{ kJ/kg}$$

By SFEE, assuming negligible kinetic and potential energy changes

$$q - w = h_2 - h_1$$

$$= -302$$

whence

$$q = w - 302$$

$$= -311 - 302 = -613 \text{ kJ/kg}$$

Example 7.2 Application of Clapeyron's equation to calculate the vapour pressure of a liquid. Clapeyron's equation can be used to calculate the approximate vapour pressure of a liquid at an arbitrary temperature in conjunction with a relation for the latent heat of a substance known as Trouton's rule*, which states that

$$\frac{\Delta \tilde{h}_v}{T_B} \approx 88 \text{ J/K mol}$$

where $\Delta \tilde{h}_v$ is the molar latent heat of vaporization of a substance and T_B is the boiling point of the same substance at 101 325 N/m².

On substituting this into Clapeyron's equation and integrating once

$$\ln \frac{p}{101\ 325} = - \frac{88 T_B}{\tilde{R}} \left(\frac{1}{T} - \frac{1}{T_B} \right)$$

from which the vapour pressure p at temperature T can be found.

As a numerical example, the vapour pressure of benzene at 303 K may be found given that the boiling point of benzene at 1 atmospheric pressure is 353 K. Assuming Trouton's rule is obeyed

$$\ln \frac{p}{101\ 325} = - \frac{88 \times 353}{8.3143} \left(\frac{1}{303} - \frac{1}{353} \right)$$

whence

$$p = 17\ 650 \text{ N/m}^2$$

It should be borne in mind that the latent heat of vaporization does in fact vary with temperature and pressure and that the above method yields only an approximation to the actual value.

PROBLEMS

1. Show that the difference between the specific heat capacities of an ideal gas and a van der Waals gas are respectively

$$R \text{ and } \frac{R}{1 - 2a(v - b)^2/RTv^3}$$

(Use the relation derived in Example 3.3)

2. For a substance obeying the Clausius equation of state, show that c_p is a function of temperature only.

---

* Trouton's rule holds well for non-polar liquids of molecular weight about 100. If hydrogen-bonding occurs in the liquid, e.g. in ethanol or water, the ratio is greater than that assumed, since now the latent heat also includes the energy required to break the extra bonding forces between the molecules of the liquid.

3. Show that the internal energy and entropy per unit mass of a van der Waals gas are given by

$$u = \text{constant} + \int c_v \, dT - \frac{a}{v}$$

$$s = \text{constant} + \int \frac{c_v}{T} \, dT + R \ln (v - b)$$

4. Show that the work done by a gas obeying van der Waals' equation is given by

$$W = n\tilde{R}T \ln \left(\frac{V_2 - nb}{V_1 - nb} \right) + \frac{an^2}{V_2} - \frac{an^2}{V_1}$$

5. Show that the difference between the two principal specific heats of a van der Waals' gas is

$$\tilde{C}_p - \tilde{C}_v = \tilde{R} + \tilde{R} \left(\frac{2a}{p\tilde{v}^2} - \frac{2ab}{p\tilde{v}^3} \right) \left(1 - \frac{a}{p\tilde{v}^2} + \frac{2ab}{p\tilde{v}^3} \right)^{-1}$$

6. Show that the relation governing the behaviour of a van der Waals' gas undergoing an adiabatic reversible process is

$$T (\tilde{v} - b)^{\tilde{R}/\tilde{C}_v} = \text{constant}$$

7. Show that the Joule–Kelvin coefficient for a van der Waals' gas is given by

$$\mu_{JK} = \frac{\tilde{v}}{\tilde{C}_p} - \left\{ \frac{2a (\tilde{v} - b)^2 - \tilde{R}Tb\tilde{v}^2}{\tilde{R}T\tilde{v}^3 - 2a (\tilde{v} - b)^2} \right\}$$

8. Show that the isobaric expansivity and isothermal compressibility for a gas obeying the van der Waals equation are respectively

$$\beta_p = \left\{ T \left(1 - \frac{2a}{\tilde{R}T\tilde{v}} + \frac{pb}{\tilde{R}T} + \frac{3ab}{\tilde{R}T\tilde{v}^2} \right) \right\}^{-1}$$

$$\kappa_T = \frac{1}{p} \left(1 - \frac{b}{\tilde{v}} \right) \left(1 - \frac{2a}{p\tilde{v}^2} + \frac{2ab}{p\tilde{v}^3} \right)^{-1}$$

9. Show that the fugacity of a gas obeying the van der Waals' equation of state is given by

$$\ln f = \ln \left(\frac{\tilde{R}T}{\tilde{v} - b} \right) + \frac{b}{\tilde{v} - b} - \frac{2a}{\tilde{R}T\tilde{v}}$$

10. (a) Show that for large values of v, van der Waals equation may be written as

$$pv = RT + \frac{RTb - a}{v} + \frac{RTb^2}{v^2} \left(1 + \frac{b}{v} + \frac{b^2}{v^2} + \dots \right)$$

(This can be obtained by direct expansion in terms of $1/v$), and
(b) that for small values of p, it may be written as

$$pv = RT + p \left(b - \frac{a}{RT} \right) + \frac{ap^2}{R^2T^2} \left(2b - \frac{a}{RT} \right) + \dots$$

(This can be obtained by inverting the series in (a).)
11. Over a considerable range of temperature the equation of state of a solid may be represented as

$$v = v_0 (1 + 3\alpha T - \kappa p)$$

where v_0 is a constant and α, κ are respectively the coefficient of linear expansion and isothermal compressibility. Show that the internal energy is given by

$$u = u_0 + cT + \frac{1}{2\kappa} \frac{(v - v_0)^2}{v_0}$$

Also show that the entropy is given, apart from a constant, by

$$c \ln T + 3\alpha (v - v_0)/\kappa$$

and that

$$c_v = c$$

$$c_p = c + 9\alpha^2 T v_0/\kappa$$

12. Derive expressions for the slopes of constant pressure and constant volume lines on a T–s diagram for a perfect gas and show that the volume line is steeper.
13. What is the law of corresponding states? What is its basis? In what respect is the law useful? The critical temperature and pressure of benzene are 561.7 K and 47.9 atm respectively, and those of chloroform are 533.2 K and 54.9 atm; benzene boils at 353.2 K. What would you expect the vapour pressure of chloroform to be at 335.2 K?

CHAPTER EIGHT

Thermodynamics of Multi-component Systems

8.1 THE PHASE RULE

Before proceeding to our subject proper, a phase of a system will first be defined, although this term has already been used on several occasions before. By a phase of a system it is meant that collection of all homogeneous parts of the system having the same intensive state. Thus, a phase is uniform in its intensive state as well as in composition. The condition of homogeneity or uniformity, as required by the definition, is rather demanding. It requires on the one hand that a phase must be so large that departure from homogeneity due to capillarity at its boundary with other neighbouring phases is negligible. On the other hand, it also requires that a phase should not be so large that gravitational forces between parts of the phase will set up pressure and density gradients so that these intensive properties are no longer uniform.

Intensive phase properties, such as pressure, temperature, specific volume, mole fraction, etc., are by definition independent of the size of the phase. They are not, however, all independent of each other. The number of independent intensive phase properties for an equilibrium state depends on the number of phases present and the composition of the system expressed in terms of the number of its components. The number of components, n, is defined as the minimum number of independently variable chemical species necessary to describe the composition of all the phases in the system. In the absence of chemical reactions and semi-permeable membranes, the number of components of a phase in a multiphase system is identical with the number of molecular species present. Chemical reactions reduce the number of components by the number of independently allowed reactions. The number of components for the system is therefore the union of all the sets of components of the different phases present in the system—provided that no species in this set can be formed from species present in the subsets, by mixing or chemical reaction.

Phase rule for open systems

A closed simple system has, according to the state principle, two independent intensive properties. When this system is open to transfer of masses of n different components, the masses of these n components may be altered independently of each other and of the intensive states of the system. Thus, the number of independent properties for an open simple system having n components is $(n + 2)$.

On the other hand, let F denote the number of independent intensive phase properties of this same open simple system. When all independent intensive phase properties are fixed, the intensive state of each phase is fixed; however, the size of the phase is not. For an open system, the size of each phase is independent of the size of every other phase, as well as of the intensive state of each phase. It follows, therefore, that the number of independent properties of the entire open simple system is given by $F + P$, where P is the number of phases present. Alternatively, it may be said that the F intensive properties fix the intensive states of the phases and the P masses of each phase fix the size of the system.

Equating the two expressions, we have

$$F + P = n + 2 \qquad (8.1)$$

Phase rule for closed systems

Contrary to an open system, in a closed system no possible variation in the phase masses can alter all of them in the same direction. Let the number of phase masses that can be independently varied while the intensive phase properties are held constant be r; then the number of independent properties for the closed simple system is $F + r$, where F is the number of independent intensive phase properties. According to the state principle, a closed simple system has only two independent properties, therefore

$$F + r = 2 \qquad (8.2)$$

Now the condition that has to be satisfied for an allowed variation in the phase masses is that the total change in mass of each component in the system is zero. Thus, for any change in which the intensive phase properties are held constant

$$
\begin{aligned}
x_1^1 \, \Delta m^1 + x_1^2 \, \Delta m^2 + \ldots + x_1^P \, \Delta m^P &= 0 \\
 \vdots \qquad\qquad\qquad \vdots \\
x_n^1 \, \Delta m^1 + x_n^2 \, \Delta m^2 + \ldots + x_n^P \, \Delta m^P &= 0
\end{aligned}
\qquad (8.3)
$$

where Δm^1, Δm^2, ..., Δm^P denote the variation in the masses of the P phases and x_j^i the mass fraction of component j in phase i.

The x_j^i's can be taken to be constant for an infinitesimal change in the phase masses. The above set of equations therefore represents a system of n homogeneous linear equations in P unknowns. For this system to be consistent for $P \geq n$, $P - n$ of the unknowns may be chosen arbitrarily. Thus, of the P phase masses, $P - n$ are independent, therefore

$$r = P - n \qquad (8.4)$$

Hence

$$F + P - n = 2$$

or

$$F + P = n + 2 \qquad (8.1)$$

This is the same expression as obtained for the open simple system, only now the additional restriction $P \geq n$ applies. This is because r cannot be less than zero.

8.2 PARTIAL PROPERTIES

Consider a single-phase system consisting of n components; the number of independent phase properties necessary for defining the state of the phase is equal to $n + 2$. Thus, for any extensive property, V say, we have

$$V = V(p, T, m_1, m_2, ..., m_n) \qquad (8.5)$$

where the m's represent the masses of each component present in the system.

For a general variation in V in which all the variables undergo changes, by differentiating the above relation

$$dV = \left(\frac{\partial V}{\partial p}\right)_{T,m} dp + \left(\frac{\partial V}{\partial T}\right)_{p,m} dT + \left(\frac{\partial V}{\partial m_1}\right)_{p,T,m} dm_1$$

$$+ \left(\frac{\partial V}{\partial m_2}\right)_{p,T,m} dm_2 + ... \qquad (8.6)$$

where the subscript m on a differential coefficient indicates that all m's are to be held constant except the one appearing in the differential coefficient itself.

In particular, if we have a change involving addition of masses to the system at constant p and T, equation (8.6) becomes

$$dV = v_1 dm_1 + v_2 dm_2 + ... = \Sigma v_i dm_i \qquad (8.7)$$

where

$$v_i = \left(\frac{\partial V}{\partial m_i}\right)_{p,T,m}$$

The above relation is analogous to that for an open one-component system for which the addition of mass dm will cause a change in the volume of the system by an amount

$$dV = v\, dm \tag{8.8}$$

where v is the specific volume of the system at the given p and T.

The quantities v_i are thus seen to be analogous to the specific volume for a single component system. They are known as partial volumes of i in the phase of which i is a component. Like v, they are intensive properties whose values are determined by the temperature and pressure of the phase.

Now a phase of any composition may be increased in size without change in its intensive properties by holding p and T constant and without change in its composition by maintaining the original mass fraction for each addition of masses. Thus

$$\frac{dm_1}{m_1} = \frac{dm_2}{m_2} = \ldots = k \tag{8.9}$$

To each stage of this process, equation (8.7) applies for the change in volume of the system. Since each partial volume remains constant throughout the process, this may be integrated to give

$$V = v_1 m_1 + v_2 m_2 + \ldots \tag{8.10}$$

This, of course, is an entirely general relation between any extensive property and its corresponding partial properties of a phase. Note that it merely re-affirms the rule that extensive properties are additive.

A relation between partial properties that does not involve the corresponding extensive property may now be derived. On differentiating, equation (8.10) gives

$$dV = v_1\, dm_1 + m_1\, dv_1 + v_2\, dm_2 + m_2\, dv_2 + \ldots \tag{8.11}$$

On subtracting equation (8.7) from this

$$m_1\, dv_1 + m_2\, dv_2 + \ldots = \Sigma\, m_i\, dv_i = 0 \tag{8.12}$$

This is another general relation among partial properties, for changes at constant p and T.

So far, we have discussed partial properties in terms of a derivative with respect to masses of the components. Analogous relations to the above equations exist for the molal quantities, which are partial derivatives with respect to the number of moles of each substance in the system. In fact, using molal quantities gives in general simpler relations for multicomponent systems, particularly where perfect gases or chemical reactions are involved.

It can be shown that thermodynamic potential functions which have been defined for a single component system also apply to a multicomponent system for the partial properties. For example, consider the Gibbs function

$$G = H - TS$$

For a small variation, this gives

$$dG = dH - TdS - SdT \qquad (8.13)$$

If this occurs at constant p and T such that all m's except m_i are also held constant, then from equation (8.6)

$$dG = \left(\frac{\partial G}{\partial m_i}\right)_{p,T,m} dm_i \qquad (8.14)$$

and similar relations for the other properties. Thus,

$$\left(\frac{\partial G}{\partial m_i}\right)_{p,T,m} = \left(\frac{\partial H}{\partial m_i}\right)_{p,T,m} - T\left(\frac{\partial S}{\partial m_i}\right)_{p,T,m}$$

i.e.

$$g_i = h_i - Ts_i \qquad (8.15)$$

Compare this relation to that for the specific properties of a single component system. Similar expressions can be derived for molal properties. Thus

$$\tilde{g}_i = \tilde{h}_i - T\tilde{s}_i \qquad (8.16)$$

where

$$\tilde{g}_i = \left(\frac{\partial G}{\partial n_i}\right)_{p,T,n}$$

etc.

8.3 GIBBS–DUHEM EQUATION

Referring to equation (8.6) in the previous section, let us consider the property of interest be G; then

$$dG = \left(\frac{\partial G}{\partial p}\right)_{T,m} dp + \left(\frac{\partial G}{\partial T}\right)_{p,m} dT + \sum \left(\frac{\partial G}{\partial m_i}\right)_{p,T,m} dm_i \qquad (8.17)$$

The first two partial derivatives on the right hand side refer to a closed system, and are equal to V and $-S$ respectively, (see Section 3.7). The remaining partial derivatives are partial properties as defined in the present chapter. Denoting

$$\mu_i = \left(\frac{\partial G}{\partial m_i}\right)_{p,T,m} \tag{8.18}$$

the above equation becomes

$$dG = V dp - S dT + \Sigma \mu_i dm_i \tag{8.19}$$

This equation is known as Gibbs equation for an open system. It may be expressed in the following alternative forms by substituting in equation (8.19) the defining relations

$$dA = d(G - pV)$$
$$dH = d(G + TS)$$
$$dU = d(G - pV + TS)$$

leading to

$$dA = - pdV - S dT + \Sigma \mu_i dm_i \tag{8.20}$$
$$dH = V dp + T dS + \Sigma \mu_i dm_i \tag{8.21}$$
$$dU = - pdV + T dS + \Sigma \mu_i dm_i \tag{8.22}$$

An inspection of equations (8.19) to (8.22) shows that

$$\mu_i = \left(\frac{\partial G}{\partial m_i}\right)_{p,T,m} = \left(\frac{\partial A}{\partial m_i}\right)_{V,T,m} = \left(\frac{\partial H}{\partial m_i}\right)_{p,S,m} = \left(\frac{\partial U}{\partial m_i}\right)_{V,S,m}$$

Only the first of these partial derivatives corresponds to a partial property— namely a partial derivative with respect to m_i with p and T held constant. The quantity μ_i is called the chemical potential. However, we shall retain our notation g_i when we are emphasizing its role as a partial property.

Referring back to equation (8.10), since G is an extensive property

$$G = \Sigma g_i m_i$$

Differentiating

$$dG = \Sigma g_i dm_i + \Sigma m_i dg_i$$

Comparing this with equation (8.19), it is seen that

$$V dp - S dT - \Sigma m_i dg_i = 0 \tag{8.23}$$

This is known as the Gibbs–Duhem equation. It is applicable to any homogeneous phase. When T and p remain constant, this reduces to

$$\Sigma \, m_i \, dg_i \;=\; 0 \tag{8.24}$$

As already shown, this special case of the Gibbs–Duhem equation holds for any arbitrary property.

8.4 EQUILIBRIUM WITH RESPECT TO TRANSFER OF COMPONENTS

For a simple system consisting of n components and P phases, it is necessary that at equilibrium the pressure and temperature should be uniform throughout (see Section 3.8) and a sufficient criterion for equilibrium at constant p and T is

$$dG_{p,T} \;=\; 0 \tag{3.47}$$

thus

$$\sum_{j=1}^{n} \; \sum_{i=1}^{P} \mu_j^i \, dm_j^i \;=\; 0 \tag{8.25}$$

Also, since there is no addition of masses from external cources

$$\sum_{i=1}^{P} \; dm_j^i \;=\; 0 \qquad j \;=\; 1, 2, ..., n \tag{8.26}$$

Let us consider the special case in which only component 1 is allowed to vary; then the above relations become

$$\mu_1^1 \, dm_1^1 + \mu_1^2 \, dm_1^2 + ... + \mu_1^P dm_1^P \;=\; 0 \tag{8.27}$$

and

$$dm_1^1 + dm_1^2 + ... + dm_1^P \;=\; 0 \tag{8.28}$$

Eliminating dm_1^1 between equations (8.27) and (8.28)

$$\mu_1^1 \, (-dm_1^2 - dm_1^3 - ... - dm_1^P) + \mu_1^2 \, dm_1^2 + ... + \mu_1^P dm_1^P \;=\; 0 \tag{8.29}$$

or

$$(\mu_1^2 - \mu_1^1) \, dm_1^2 + ... + (\mu_1^P - \mu_1^1) \, dm_1^P \;=\; 0 \tag{8.30}$$

In order that the above equation holds for all arbitrary values of dm, it is required that each of the coefficients be zero; thus

$$\mu_1^1 \;=\; \mu_1^2 \;=\; ... \;=\; \mu_1^P \tag{8.31}$$

In words, this states that it is necessary for equilibrium of a heterogeneous system that the value of the chemical potential for each component should be the same in every phase in which the component is actually present.

8.5 CHEMICAL POTENTIAL AS AN ESCAPING TENDENCY

Let us consider here a problem akin to the previous one. Suppose now that component 1 is not actually present in a phase but present in all adjoining phases and is not prevented from entering the first phase, denoted here by the superscript a. The criterion for an allowed change of state in which dm_1^a is passed from the adjoining phases into phase a is

$$\mu_1^a \, dm_1^a + \mu_1^2 \, dm_1^2 + \ldots + \mu_1^P \, dm_1^P \geq 0 \tag{8.32}$$

and

$$dm_1^a + dm_1^2 + \ldots + dm_1^P = 0 \tag{8.33}$$

On substituting dm_1^a from (8.32) into (8.33)

$$(\mu_1^2 - \mu_1^a) \, dm_1^2 + \ldots + (\mu_1^P - \mu_1^a) \, dm_1^P \geq 0 \tag{8.34}$$

By symmetry, all the $(\mu_1^i - \mu_1^a)$ terms should be of the same sign, and since all dm_1^is are negative quantities, it is therefore necessary that

$$\mu_1^i - \mu_1^a \leq 0$$

or

$$\mu_1^a \geq \mu_1^i \tag{8.35}$$

Thus, in general, if a component j is absent from a phase but is not prevented from entering it, the relation

$$\mu_j^a \geq \mu_j^i \tag{8.36}$$

holds. It is clear that since the equality is valid for the case in which component j is actually present in all phases and the inequality holds when j is absent from a phase, the quantity μ_j^i of component j which is allowed in phase i cannot be less than that of the same component in all the adjoining phases. In this sense, the quantity μ_j can be interpreted as an escaping tendency. The value of μ_j in a phase for which j is neither an actual nor a possible component is indeterminate.

8.6 CHEMICAL POTENTIAL OF IDEAL AND REAL GASES

It has been observed in Section 8.2 that the value of a partial property of a pure phase is identical to that of the specific property. Thus the chemical potential of a pure or single-component phase is simply the specific Gibbs free energy of the phase. Consider the relation

$$dg = -s dT + v dp \tag{3.39}$$

At constant T

$$dg = v dp_T \tag{8.37}$$

For an ideal gas, this becomes

$$dg_T = \frac{RT}{p} dp_T = RT d (\ln p)_T$$

thus

$$\mu = g = RT \ln p + F(T) \tag{8.38}$$

Now if we define the standard state of an ideal gas as the state of the pure gas at 1 atm pressure (101.325 kN/m^2) and at the temperature under consideration, then

$$\mu^o = F(T) \tag{8.39}$$

Hence, the value of the chemical potential at its standard state is a function of the temperature only. Thus

$$\mu - \mu_o = RT \ln p \tag{8.40}$$

where p is the pressure in atmospheres. When other units of pressure are used, equation (8.40) should be written in the form involving the reference pressure also. This is

$$\mu - \mu_o = RT \ln \frac{p}{p_o} \tag{8.41}$$

It is obvious that equation (8.41) does not apply to real gases. However, in order to preserve the extreme simplicity of the form of the expression for the chemical potential for an isothermal process, define

$$d\mu_T = RT d (\ln f)_T \tag{8.42}$$

On integrating, this gives

$$\mu - \mu_o = RT \ln \frac{f}{f_o} \tag{8.43}$$

The parameter f defined above is known as the fugacity of the system. Like pressure, it is a property of the system. The relation of fugacity to pressure can be obtained by substituting for v in equation (8.37) using the real gas equation in terms of the compressibility factor, i.e. $pv = zRT$, whence

$$d\mu_T = \frac{zRT}{p} dp_T = zRT d (\ln p)_T \tag{8.44}$$

On dividing (8.42) by (8.44)

$$\frac{d(\ln f)_T}{d(\ln p)_T} = z$$

i.e.

$$\left.\frac{\partial(\ln f)}{\partial(\ln p)}\right|_T = z \qquad (8.45)$$

Since $z \to 1$ as $p \to 0$, it follows that

$$\lim_{p \to 0} \frac{f}{p} = 1 \qquad (8.46)$$

Thus, fugacity and pressure will be numerically equal at low pressures and deviate from each other only at high pressures when a real gas deviates from ideal behaviour. In this sense, fugacity may be regarded as a pseudo-pressure.

8.7 EFFECTS OF PRESSURE AND TEMPERATURE ON FUGACITY

Differentiating equation (8.42) with respect to pressure at constant T

$$\left(\frac{\partial \mu}{\partial p}\right)_T = RT \left(\frac{\partial \ln f}{\partial p}\right)_T$$

But for a pure gas, by equation (3.44)

$$\left(\frac{\partial \mu}{\partial p}\right)_T = \left(\frac{\partial g}{\partial p}\right)_T = v$$

Hence, the pressure dependence of f can be found through the relation

$$\left(\frac{\partial \ln f}{\partial p}\right)_T = \frac{v}{RT} \qquad (8.48)$$

To find the effect of temperature on fugacity, it is easier to begin with equation (8.43) rewritten in the form

$$\ln f - \ln f^0 = \frac{1}{R}\left(\frac{\mu}{T} - \frac{\mu^0}{T}\right)$$

On differentiating with respect to T at constant p, we have

$$\left(\frac{\partial \ln f}{\partial T}\right)_p - \left(\frac{\partial \ln f^0}{\partial T}\right)_p = \frac{1}{R}\left\{\left(\frac{\partial(\mu/T)}{\partial T}\right)_p - \left(\frac{\partial(\mu^0/T)}{\partial T}\right)_p\right\} \qquad (8.49)$$

Now, since f^0 is the fugacity of the gas in the standard state, which by definition is independent of temperature, the above equation becomes

$$\left(\frac{\partial \ln f}{\partial T}\right)_p = -\frac{1}{R}\left\{\left(-\frac{\mu}{T^2} + \frac{1}{T}\left(\frac{\partial \mu}{\partial T}\right)_p\right) - \left(-\frac{\mu^0}{T^2} + \frac{1}{T}\left(\frac{\partial \mu^0}{\partial T}\right)_p\right)\right\} \quad (8.50)$$

For a pure gas, using equation (3.45) the above equation becomes

$$\left(\frac{\partial \ln f}{\partial T}\right)_p = -\frac{h - h^0}{RT^2} \quad (8.51)$$

Thus the temperature dependence of f can be found through equation (8.51).

8.8 DEVELOPING FUGACITY VALUES FROM EXPERIMENTAL DATA

Since fugacity is a property, its values can be developed by methods similar to those described in Section 7.5. These will be briefly reviewed below.

In the presence of p–v–T data, an equation of state can be fitted for a certain range over which the fugacity of the gas can be found by direct integration of equation (8.48). The arbitrary function of T can then be evaluated by imposing the condition $f \to p$ as $p \to 0$. In the absence of p–v–T data, the following procedure can be adopted.

Subtracting equation (8.44) from (8.42)

$$\{d(\ln f) - z d(\ln p)\}_T = 0 \quad (8.52)$$

Rearranging, this becomes

$$d(\ln f)_T - d(\ln p)_T = (z - 1) d(\ln p)_T$$

i.e.

$$d \ln \frac{f}{p} = (z - 1) \left.\frac{dp}{p}\right|_T = (z - 1) \left.\frac{dp^*}{p^*}\right|_{T^*} \quad (8.53)$$

Integrating from $p^* = 0$ to p^*, and since $f/p \to 1$ as $p \to 0$, it follows that

$$\ln \frac{f}{p} = \int_0^{p^*} \left.\frac{(z - 1) dp^*}{p^*}\right|_{T^*} \quad (8.54)$$

Thus, by plotting $(z - 1)/p^*$ *versus* p^* at constant T^*, f/p may be evaluated. The graphical presentation of this ratio against p^* is known as the generalized fugacity chart.

EXAMPLES

Example 8.1 For a binary solution, show how the partial quantities can be obtained from experimental data plotted as a function of composition.

Let us consider the case involving the partial volume of a binary solution. The experimentally measured volume when plotted against composition may appear as shown in Figure 8.1.

Since

$$v = x_A v_A + x_B v_B \qquad \text{(a)}$$

and

$$x_A + x_B = 1$$

then

$$\left(\frac{\partial v}{\partial x_A}\right)_{p,T,x_B} = v_A - v_B$$

Therefore

$$x_B \left(\frac{\partial v}{\partial x_A}\right)_{p,T,x_B} = x_B (v_A - v_B) \qquad \text{(b)}$$

(a) + (b) yields

$$v + x_B \left(\frac{\partial v}{\partial x_A}\right)_{p,T,x_B} = (x_A + x_B) v_A = v_A$$

whence

$$v_A = v + (1 - x_A) \left(\frac{\partial v}{\partial x_A}\right)_{p,T,x_B}$$

Thus, at any arbitrary composition, the partial volume for A (v_A) is given by the intercept of the tangent at x_A on the pure A ordinate.

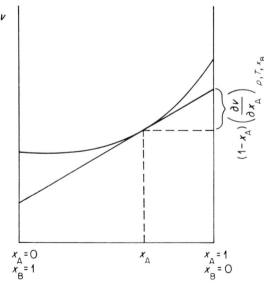

Figure 8.1

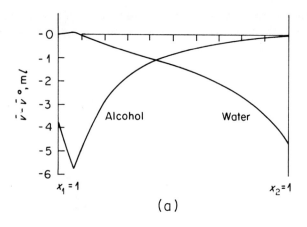

(a)

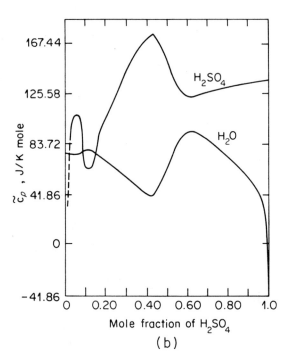

(b)

Figure 8.2 (a) Partial molal volumes of water and alcohol in a binary water–alcohol system.

(b) Partial molal heat capacities of H_3O and H_2SO_4 in aqueous sulphuric acid at 288 K (measured by Lewis and Randall). (From *Thermodynamics* by Lewis and Randall. Copyright 1961 by McGraw–Hill, Inc. Used with permission of McGraw–Hill Book Co.)

For the case of an ideal solution, we see from Example 9.1 that the volume–composition diagram is a straight line, hence the partial volume is independent of composition.

The above method and conclusion apply also to partial enthalpy of the solution.

Example 8.2 For a binary solution, discuss the behaviour of the corresponding partial properties of the two components forming the solution.

From equation (8.12)

$$\left(\frac{dv_1}{dv_2}\right)_{p,T} = -\frac{m_2}{m_1} = -\frac{x_2}{x_1}$$

It is therefore observed that v_1 decreases as v_2 increases during a certain change in composition at constant p and T. Consequently, v_1 is a maximum for the composition for which v_2 has a minimum and vice versa. For the pure state of 1, i.e. for $x_2 = 0$, the rate of change of v_2 is infinite if the rate of change of v_1 remains finite, and the rate of change of v_2 can be finite only if the rate of change of v_1 becomes zero. Figure 8.2 illustrates these characteristics for the water–alcohol as well as the aqueous sulphuric acid systems.

Example 8.3 Find the fugacity of water at 298 K and 10^7 N/m^2.

At 298 K, the vapour pressure of water is 3160 N/m^2 and the density of water is 997.0 kg/m^3; its molecular weight is 18.02.

Since the vapour pressure is low we may take $f = p = 3160$ N/m^2 at 298 K. Thus, the fugacity of the liquid at 298 K is 3160 N/m^2.

The fugacity of water at 298 K and 10^7 N/m^2 can be obtained by integrating equation (8.48), which yields

$$RT \ln \frac{f_2}{f_1} = v(p_2 - p_1)$$

On substitution

$$\ln \frac{f_2}{f_1} = \frac{18.02}{8.3143 \times 298 \times 997.0 \times 1000} (10^7 - 3160) = 0.073$$

whence

$$f_2 = 1.076 \times 3160 = 3400 \text{ N/m}^2$$

PROBLEMS

1. Show that if the pressures $p_1, p_2, \ldots, p_n$ of the separate n phases of a thermo-dynamic system are not equal (because of surface tension, membranes, etc), and that no external work is done other than that by the pressures, then for a reversible isothermal change in which the pressures are kept constant

$$\sum_{i=1}^{m} g_i dm_i = 0$$

where g_r is evaluated at temperature T and pressure p_r.

2. How must the phase rule be modified if all the components are not assumed present in each phase? For example, in a system comprising an aqueous NaCl solution and its vapour, NaCl would not usually be considered to be present in the gaseous phase.

CHAPTER NINE

Properties of Ideal Solutions

9.1 THE CONCEPT OF A SOLUTION

A solution is any *homogenous* system which consists of two or more chemically pure substances, whether it is solid, liquid or gaseous. Solutions must, therefore, be distinguished from chemical compounds, on one hand, and mixtures (in the general sense), on the other. A chemical compound consists of molecules of only one kind, the masses of the constituents of which it is formed standing in a definite ratio. It is quite different from a solution, in which the molecules or atoms of the constituents can assume any arbitrary ratio; within certain limits. From the point of the phase rule, the former is a system of one component while the latter must be regarded as a homogeneous system with several components.

The difference between solutions and mixtures is not so easily distinguishable. However, the subtlety involved is clearly brought out in the case of solid solu-tions. We know, at present, that solids exist mostly in the crystalling state, and we say that two substances are mutually soluble, in the solid state, when tions. We know, at present, that solids exist mostly in the crystalline state, when they form a joint crystalline lattice, or a mixed crystal in which the lattice points are occupied, in part, by atoms or molecules of one substance and in party by the other. All other cases, viz. those in which the individual constituents do not form a joint crystal but exists as pure-constituent crystals side by side, must be classed as mixtures. Thus, unlike a solution, a mixture is a *heterogeneous* system consisting of two or more coexisting phases.

The conditions in a liquid solution are similar. A liquid can, in general, dissolve other liquids, solids or gases, the true solution being homogeneous and characterized by a uniform distribution over its whole volume of the mole-cules or atoms of all constituents. On the other hand, liquid mixtures, such as suspensions, emulsions and colloidal solutions, are, like the solid mixtures, heterogeneous and represent two or more phases intermixed to a finer or coarser degree of dispersion.

In the case of gases, the difference between a solution and a mixture dis-appears, since all gases are mutually soluble, forming a homogeneous system in which the molecules of all constituents are uniformly distributed.

The process of forming a solution is often accompanied by thermal and mechanical effects which show that there is a certain degree of interaction between the molecules of the different constituents. For example, when ethyl alcohol and water are dissolved in equal parts, there is a considerable contraction in volume—the volume of the solution being about 5 percent smaller than the sum of the volumes of the unmixed components. At the same time, a considerable amount of heat is developed. Both effects are characteristic of the general behaviour when a solution is formed. However, for gases in general and for certain pairs of liquids whose constituents bear close chemical resemblance, so that interaction between atoms of different kinds is nearly the same as that between atoms of the same kind, very little diminution in volume or heating occurs in forming solutions. As the solution process usually takes place at constant pressure, the heat of solution is a measure of the enthalpy change of the system before and after mixing. We can, therefore, imagine as an ideal case, a solution in which the volumes, as well as the enthalpies, are strictly additive and no change occurs in these quantities when a solution is formed. Such solutions have been called ideal solutions by Lewis. Since the constancies in volume and enthalpy are consequences of each other (as we shall later show), we need not specify both in our definition of an ideal solution. Moreover, since the partial properties have to be specified by the temperature and pressure of the multicomponent system, an ideal solution will be defined as follows.

An ideal solution is one such that the partial molal volume of each component in solution is the same as its molal volume when it exists by itself at the temperature and pressure of the solution.

Mathematically, this can be written as

$$\tilde{v}_i - \tilde{v}'_i = 0 \qquad (9.1)$$

where $\tilde{v}_i$ is the partial molal volume of constituent i in a solution and $\tilde{v}'_i$ is the molal volume of constituent i in its pure state at the same temperature and pressure as the solution

9.2 AMAGAT–LEDUC LAW OF ADDICTIVE VOLUMES

A direct consequence of the definition of an ideal solution is that it obeys the Amagat–Leduc law of additive volumes. This can be easily derived as follows. Since

$$\tilde{v}_i - \tilde{v}'_i = 0$$

we have

$$\Sigma n_i (\tilde{v}_i - \tilde{v}'_i) = 0$$

hence

$$\Sigma n_i \tilde{v}_i = \Sigma n_i \tilde{v}'_i = \Sigma V'_i$$

where V'_i is the volume of component i existing in its pure state at the same temperature and pressure as the solution. But, from our earlier discussion on partial properties, the total volume of the solution is given by

$$V = \Sigma n_i \tilde{v}_i$$

Hence

$$V = \Sigma V'_i \tag{9.2}$$

In words, this states that the total volume of an ideal solution is equal to the sum of the individual volumes of each component when they exists alone at the temperature and pressure of the solution.

9.3 MIXTURE OF IDEAL GASES

A mixture of ideal gases is a special case of an ideal solution. The Amagat–Leduc law of additive volumes is therefore fundamental to its behaviour. From this, other experimentally obtained laws governing the behaviour of an ideal gas mixture will be derived.

When the gaseous components forming the mixture behave ideally in their pure state, the volume of component i, say, at the temperature T and pressure p of the mixture is given by

$$pV'_i = n_i RT \tag{9.3}$$

It follows therefore from Amagat–Leduc Law that

$$V = \Sigma V'_i = \sum \frac{n_i RT}{p} = \frac{RT}{p} \Sigma n_i$$

By virtue of the fact that matter is conserved,

$$\Sigma n_i = n \tag{9.4}$$

therefore

$$pV = nRT \tag{9.5}$$

Thus a mixture of ideal gases is itself an ideal gas.

Next, consider each component to occupy the entire volume of the mixture, and let the pressure of component i when its temperature is the same as that of the mixture be p_i; then

$$p_i V = n_i RT$$

Comparing this with equation (9.3)

$$p_i V = p V'_i \tag{9.6}$$

Summing over all i's in the mixture, this gives

$$V \Sigma p_i = p \Sigma V'_i$$

which, by virtue of Amagat–Leduc Law, becomes

$$p = \Sigma p_i \tag{9.7}$$

Equation (9.7) is known as Dalton's law of additive pressures.
Now, from equation (9.6)

$$\frac{V'_i}{V} = \frac{p_i}{p}$$

But from equations (9.3) and (9.5)

$$\frac{V'_i}{V} = \frac{n_i}{n}$$

Defining the mole fraction of component i in the mixture by $y_i = n_i/n$

$$\frac{V'_i}{V} = \frac{p_i}{p} = y_i \tag{9.8}$$

Since an ideal gas mixture behaves as an ideal gas, we can deduce the apparent molecular weight of the mixture as follows.
Since for each component

$$n_i = \frac{m_i}{M_i}$$

where m denotes the mass and M the molecular weight of an ideal gas, then

$$\Sigma\, n_i\, M_i = \Sigma\, m_i = m$$

where unsubscripted quantities refer to those of the mixture. But, for the mixture

$$n = \frac{m}{M}$$

whence

$$nM = \Sigma\, n_i\, M_i$$

or

$$M = \Sigma\, y_i\, M_i \tag{9.9}$$

When working in terms of specific properties, it is possible to define a specific gas constant for the mixture. This can be shown to be related to the specific gas constants of the components as follows.

Since

$$p = \Sigma\, p_i = \sum \frac{m_i\, R_i\, T}{V}$$

and

$$p = \frac{mRT}{V}$$

then

$$R = \sum \frac{m_i}{m} R_i = \Sigma \, w_i \, R_i \qquad (9.10)$$

where w_i denotes the mass fraction of component i in the mixture.

It is now possible to show how other properties for the mixture can be computed. However, the compatibility between ideal gas behaviour and the above definition of an ideal solution, will first be demonstrated.

Assuming ideal gas behaviour

$$\tilde{v}_i = \frac{\tilde{R}T}{p}$$

Similarly, for the pure state existing at the temperature and pressure of the mixture

$$\tilde{v}_i' = \frac{\tilde{R}T}{p}$$

therefore

$$\tilde{v}_i - \tilde{v}_i' = 0$$

Now, from equation (8.38), the partial Gibbs free energy at constant temperature for an ideal gas is

$$d\tilde{g}_i = \tilde{R}T \, d \, (\ln p_i)$$

Integrating between the state of the mixture and that of the pure phase yields

$$\tilde{g}_i - \tilde{g}_i' = \tilde{R}T \ln \frac{p_i}{p} = \tilde{R}T \ln y_i$$

Differentiating with respect of T

$$\frac{\partial \tilde{g}_i}{\partial T} - \frac{\partial \tilde{g}_i'}{\partial T} = \tilde{R} \ln y_i$$

i.e.

$$- \tilde{s}_i + \tilde{s}_i' = \tilde{R} \ln y_i \qquad (9.11)$$

Since

$$h = g + Ts$$

it follows that

$$\tilde{h}_i - \tilde{h}_i' = \tilde{g}_i - \tilde{g}_i' + T \, (\tilde{s}_i - \tilde{s}_i')$$

$$= \tilde{R}T \ln y_i - T \, (\tilde{R} \ln y_i) = 0$$

Thus the enthalpy of the mixture is equal to the sum of the enthalpies of the constituents, each existing at the temperature and pressure of the mixture.

Similarly, using the relation $u = h - pv$, it can also be shown that there is no change in the internal energy on mixing.

However, since mixing is a spontaneous process, the entropy of the mixture is always greater than the entropies of the constituents before mixing when each are at the same temperature and pressure as that of the mixture. Thus, quantities involving entropy in their definitions, such as the Helmholtz and Gibbs free energies, will also indicate corresponding changes due to the change in the entropy on mixing.

From equation (9.11), the change in entropy for each component is

$$\Delta S_i = - \tilde{R} n_i \ln y_i$$

Hence, the total change in entropy for all constituents in the mixture is

$$\Delta S = - \tilde{R} \Sigma n_i \ln y_i$$

The molal entropy change for the mixing process is therefore

$$\Delta \tilde{s} = - \tilde{R} \Sigma y_i \ln y_i \tag{9.12}$$

Alternatively, the final entropy of the mixture is given by

$$\tilde{s} = \Sigma y_i \tilde{s}'_i + \tilde{R} \sum y_i \ln \frac{1}{y_i} \tag{9.13}$$

where $\tilde{s}'_i$ is to be evaluated at the temperature and pressure of the mixture.

It is seen that the entropy increase in the mixing process depends only on the number of moles of the components and not on the nature of the gases. This has given rise to what is known as Gibbs' Paradox, because when like gases mix, there is no increase in entropy. The explanation lies in the fact that when it is impossible to distinguish between the gases, the process is a reversible one since now there is no need to use any semi-permeable membranes to separate the mixture back into its components.

For enthalpy and internal energy, since there is no difference between the partial molal quantity and the molal quantity in its pure state at the temperature and pressure of the mixture, it therefore follows that

$$\tilde{h} = \Sigma y_i \tilde{h}'_i \quad \text{and} \quad \tilde{u} = \Sigma y_i \tilde{u}'_i \tag{9.14}$$

From these relations, it readily follows that the molal heat capacities of the mixture are related to those of the components by the expressions

$$\tilde{C}_p = \Sigma y_i \tilde{C}'_{v_i} \quad \text{and} \quad \tilde{C}_v = \Sigma y_i \tilde{C}'_{v_i} \tag{9.15}$$

All the above formulae, of the form

$$\tilde{x} = \Sigma y_i \tilde{x}'_i$$

involving partial molal quantities and mole fractions, can be easily converted into their counterparts involving specific quantities and mass fractions of the form

$$x = \Sigma w_i x'_i \tag{9.16}$$

and the corresponding term involving entropy change on mixing becomes

$$\Delta s = \sum w_i R_i \ln \frac{V}{V_i'} \qquad (9.17)$$

9.4 MIXTURE OF AN IDEAL GAS AND A CONDENSABLE VAPOUR: PSYCHROMETRY

Before proceeding to more complicated problems on gaseous mixtures and ideal solutions in general, the problem of a mixture of ideal gases that is in contact with a solid or liquid phase of one of its components will first be considered. This problem has important engineering applications because many such processes involve air-streams which invariably carry with them a certain amount of water vapour. The following discussion will be devoted exclusively to the study of the behaviour of a mixture of air and water vapour. It should be realized, however, that the methods developed are equally applicable to any other combination of appropriate substances. The following definitions will first be introduced.

Specific humidity ω is the ratio of the mass of water vapour in a given volume of the mixture to the mass of air in the same volume. Thus

$$\omega = \frac{m_s}{m_a} = \frac{V}{m_a}\frac{m_s}{V} = \frac{v_a}{v_s} \qquad (9.18)$$

where the subscripts s and a refer to steam (water vapour) and air respectively. *Relative humidity* ϕ is the ratio of the density of water vapour to the density of saturated water vapour at the temperature of the mixture. Thus

$$\phi = \frac{\rho_s}{\rho_{sat}} = \frac{v_{sat}}{v_s} \qquad (9.19)$$

When $\phi = 1$, the mixture is said to be saturated.

The dew point of a mixture is the state to which the mixture must be cooled at constant pressure before liquid water will form.

Combining the definitions for ω and ϕ we obtain

$$\phi = \frac{v_{sat}}{v_a}\frac{v_a}{v_s} = \omega\frac{v_{sat}}{v_a} \qquad (9.20)$$

Introducing the ideal gas law into the definition of ω

$$\omega = \frac{p_s V/R_s T}{p_a V/R_a T} = \frac{R_a}{R_s} \times \frac{p_s}{p_a} = \frac{\tilde{R}/M_a}{\tilde{R}/M_s} \times \frac{p_s}{p_a} = \frac{M_s p_s}{M_a p_a} \qquad (9.21)$$

where M_s and M_a denote the molecular weights of water and air respectively. Therefore

$$\omega = 0.622 \frac{p_s}{p_a}$$ (9.22)

Likewise, ϕ can be written as

$$\phi = \frac{p_s}{p_{sat}} = \frac{p_s}{p_a} \frac{p_a}{p_{sat}} = \frac{\omega}{0.622} \frac{p_a}{p_{sat}}$$ (9.23)

By Dalton's law of additive pressures, the atmospheric pressure is given by

$$p = p_a + p_s$$ (9.24)

If a gaseous mixture of air and water vapour is cooled at constant pressure, the partial pressure of each component will remain constant so long as the dew point has not been reached. This is evident from the fact that $y_i = p_i/p$ remains constant prior to condensation. Below the dew point, condensation of the vapour to the liquid phase will alter the composition in the gaseous phase and hence the partial pressure of the components.

Notice that if the liquid phase is present, it is subjected to a higher pressure then its saturated vapour pressure and is therefore in a compressed state. The exact values of its properties ought to be determined by the formulae given in Section 5.3. However, since v_f is rather small, for all practical purposes, these can be assumed to be nearly the same as those of the saturated liquid at the same temperature.

Fundamental of psychrometrics

Because of the importance of air–water vapour mixtures in engineering processes, such as air-conditioning and manufacturing processes, special charts have been prepared for calculations involving this mixture. These are called psychrometric charts. Basically, this is a plot of specific humidity *versus* dry bulb temperature. Once the total pressure for which the chart is to be constructed has been fixed, lines of constant relative humidity, constant specific volume, constant wet bulb temperature and enthalpy can be plotted on the chart. These are usually prepared for standard atmospheric pressure with correction charts for other pressures. Enthalpy values are given per unit mass of dry air.

A typical plot of the psychrometric chart is given in Figure 9.1. Various processes of engineering interest are also illustrated in this figure and a summary of these processes given in Table 9.1.

9.5 MIXTURE OF DENSE GASES

Up to now, gaseous mixtures for which the components behave ideally have been discussed, where a mixture of ideal gases is itself an ideal gas with its properties given by weighting the corresponding components properties in a certain manner, with the exception of entropy and quantities involving entropy

Table 9.1   Summary of psychrometric processes

Process	Indication in Figure 9.1	Final air conditions	Equipment used
Heating	A–B	S.H. and D.P. unchanged, R.H. reduced, W.B.T. increased.	Air passes over surface air heater.
Heating with humidification	A–H	D.P. and S.H. increased. R.H. generally increased. (This process is for *winter air conditioning*.)	Air passes over surface heater with controlled injection of steam.
Cooling above dew point	A–C	S.H. and D.P. unchanged, R.H. increased, W.B.T. reduced. No condensation of vapour.	Air passes through air washer using refrige-
Cooling to dew point	A–D	S.H. and D.P. unchanged, R.H. becomes 100 percent. D.B.T. and W.B.T. coincide, i.e. the air is saturated but there is no condensation of vapour.	rated water for spray nozzles, or air passes over surface cooler.
Cooling with dehumidi- fication	A–D–E–E′	S.H. and D.P. reduced. Part of the humidity has condensed. (This process is for *summer air conditioning*.)	Air is subcooled in cooling equipment (see above two processes) to state E with controlled moisture removal. It is then passed over air heater with controlled heat input to reach desired state E′.
Cooling with humidification	A–I	D.P. and S.H. increased. R.H. increased.	Air passes over surface cooler with controlled injection of steam.
Adiabatic satu- ration (humidi- fication)	A–F	This is a special boundary case between A–H and A–I having zero heat input. W.B.T. un- changed; R.H., S.H. and D.P. increased.	Air passes through air washer using recircu- lated water for spray nozzles.
Adiabatic mixing	A–M–G–	Final mixture condition M is intermediate between initial air conditions A and G. M can be obtained by joining A and G with a straight line and then subdividing this in inverse pro- portion to the masses of A and G.	Mixing chamber

Key   D.P.   = Dew point
D.B.T. = Dry bulb temperature
W.B.T. = Wet bulb temperature
S.H.   = Specific humidity
R.H.   = Relative humidity

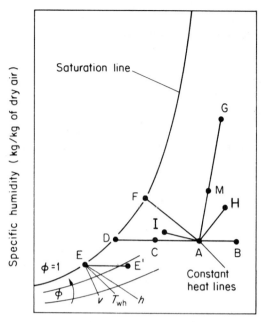

Dry bulb temperature (K)

Figure 9.1

in their definitions. As the ideal gas law is not satisfactory for dense gases, various other techniques have been developed for treating mixtures of real gases. We shall first consider below the extension of van der Waals' equation to a mixture.

The basis for treating ideal solutions is the Amagat–Leduc law of additive volumes. But, since van der Waals' equation gives a cubic in $\tilde{v}$, the results of a van der Waals mixture cannot be expressed in a simple form consistent with the law of additive volumes. In the solution of a problem, the partial volumes are first guessed (say, by using the ideal gas law) and then the pressure is calculated for each gas from their respective van der Waals equation. Generally, trial and error procedures are required until the pressures calculated for each component are equal.

Because of the difficulties experienced using the law of additive volumes with van der Waals' gases, attempts have been made in using the law of additive pressures instead. Though the laws of additive volumes and pressures are equivalent for ideal gases, there is no reason to expect that the same is true for real gases also. One should note that the latter even fails to represent the mixing of a gas with itself.

Consider one mole of a mixture, of volume $\tilde{v}$. Let y_i be the mole fraction of component i in the mixture. Now van der Waals equation written for N moles of a gas is

$$\left(p + \frac{N^2 a}{V^2}\right)(V - Nb) = NRT \qquad (9.25)$$

Thus, for y_i moles of gas i occupying a volume $\tilde{v}$, this becomes

$$\left(p_i + \frac{y_i^2 a_i}{\tilde{v}^2}\right)(\tilde{v} - y_i b_i) = y_i RT \qquad (9.26)$$

or

$$p_i = \frac{y_i RT}{\tilde{v} - y_i b_i} - \frac{y_i^2 a_i}{\tilde{v}^2} \qquad (9.27)$$

By the law of additive pressures

$$p = \Sigma p_i$$

$$= RT\left(\Sigma \frac{y_i}{\tilde{v} - y_i b_i}\right) - \frac{1}{\tilde{v}^2} \Sigma a_i y_i^2 \qquad (9.28)$$

Comparing with the van der Waals equation for which

$$p = \frac{RT}{\tilde{v} - b} - \frac{a}{\tilde{v}^2}$$

leads to the result that a mixture of van der Waals gases is itself a van der Waals gas with constants given by

$$a = \Sigma a_i y_i^2 \qquad (9.29)$$

and

$$b = \tilde{v} - \frac{1}{\displaystyle\sum \frac{y_i}{\tilde{v} - y_i b_i}} \qquad (9.30)$$

It is obvious from these results that a mixture of the same gas does not result in the same van der Waals constants. Thus it is seen that the van der Waals equation of state is not very satisfactory for use with gaseous mixtures. Other equations of state and the generalized compressibility chart have therefore come into use also. As procedures for using other equations of state are similar to those given for the van der Waals equation, we shall devote our main attention to the use of the generalized compressibility chart in the following discussion.

Consider n_i moles of gas i of initial volume V_i', pressure p and temperature T being brought together with other components at the same p and T so as to form an ideal solution. Let p_i be its partial pressure in the mixture whose final temperature remains at T. Thus

before mixing

$$pV_i' = n_i z_i \tilde{R}T \qquad (9.31)$$

after mixing

$$p_i V = n_i z_i \tilde{R} T \qquad (9.32)$$

Dividing yields

$$\frac{V_i'}{V} = \frac{p_i}{p} \qquad (9.33)$$

From this it is obvious that for real gases obeying the compressibility chart, the law of additive pressures follows as a consequence of the Amagat–Leduc law of additive volumes and the pressure of the mixture is equal to the pressure of the components before mixing when each exists at its partial volume.

It remains only therefore to evaluate the compressibility factor for the mixture so that its state can be completely determined. This will be obtained using both the laws of additive volume and pressure.

(i) Additive volumes

For the ith component initially before mixing

$$pV_i' = n_i z_i \tilde{R} T \qquad (9.31)$$

and for the mixture

$$pV = nz\tilde{R}T \qquad (9.34)$$

By the additive volume requirement,

$$V = \Sigma \, V_i' = \sum_i n_i z_i \frac{\tilde{R}T}{p} = nz \frac{\tilde{R}T}{p} \qquad (9.35)$$

hence

$$nz = \Sigma \, n_i z_i$$

i.e.

$$z = \Sigma \, y_i z_i \qquad (9.36)$$

(ii) Additive pressures

For the ith component in the mixture initially occupying the same volume as the mixture

$$p_i V = n_i z_i \tilde{R} T \qquad (9.36)$$

Hence

$$p = \Sigma \, p_i = \sum n_i z_i \frac{\tilde{R}T}{V} = nz \frac{\tilde{R}T}{V} \qquad (9.37)$$

which also gives

$$z = \Sigma \, y_i z_i \qquad (9.36)$$

Thus, once again we see the equivalence of the laws of additive volumes and pressures for gases obeying the generalized compressibility chart. These two laws taken together are therefore not sufficient to imply ideal gas behaviour. Other conditions still need to be specified as shown in the next section.

In general, the assumption that gases form ideal solutions is of much higher accuracy than the ideal gas mixture assumption. This may be applied with satisfactory accuracy to gaseous mixtures for which the reduced pressure of each component is less than 0.8.

The above procedure for real gas mixtures is no more satisfactory if the gases form non-ideal solutions. An empirical approach is to define pseudo-critical constants for the mixture and use the generalized chart to determine the compressibility factor for the mixture.

Kay has proposed a linear combination of the critical pressures and temperatures of the components weighted by their respective mole fractions, i.e. these quantities are to be evaluated from the relation

$$k_{\text{mix}} = \Sigma \, y_i k_i \qquad (9.38)$$

where k denotes either the critical pressure or temperature. This method gives results of reasonable accuracy for mixtures. Other combination rules are also in use. These are generally much more complicated, but they yield somewhat better accuracy. Two of the commonest forms are

Square-root combination

$$k_{\text{mix}} = (\Sigma \, y_i k_i^{1/2})^2 \qquad (9.39)$$

Cube-root or Lorentz combination

$$k_{\text{mix}} = \frac{1}{8} \, \Sigma_i \, \Sigma_j \, y_i y_j (k_i^{1/3} + k_j^{1/3})^3 \qquad (9.40)$$

9.6 SUFFICIENCY CONDITION FOR AN IDEAL GAS MIXTURE

It was seen earlier that the laws of additive volumes and pressures can be applied to both ideal gas mixtures as well as real gas mixtures. These two laws taken together are therefore not sufficient to guarantee ideal gas behaviour. Though for both cases the relation

$$\frac{V_i'}{V} = \frac{p_i}{p}$$

applies, it will be shown that these ratios have different meanings for ideal and real gas mixtures. For the former, recalling equation (9.8), these ratios also equal the mole fraction of component i in the mixture. For the latter, the above relation is derived for gases satisfying the equation of state $pV = nzRT$. Applying this equation to the gas before and after mixing as well as to the mixture as a whole

$$\frac{V_i'}{V} = \frac{p_i}{p} = \frac{n_i z_i}{nz} \qquad (9.41)$$

Thus, the fractional volume or pressure is no longer equal to the mole fraction of the component in the mixture. Hence, the condition that these ratios be equal to the mole fraction is therefore sufficient to guarantee ideal gas behaviour.

9.7 FUGACITY AND IDEAL SOLUTION

In Section 8.6 it was seen that the idea of fugacity arises from attempts to preserve the simplicity of the form for the change in chemical potential or Gibbs free energy for an isothermal process. For a pure component, this is defined by the relation

$$\mathrm{d}\tilde{\mu}_T = \tilde{R}T\,\mathrm{d}\,(\ln \tilde{f})_T \qquad (8.42)$$

If now we consider a solution of gases existing in equilibrium with its pure constituents through a series of semi-permeable membranes, then the chemical potentials of the pure phases alone and in solution must be the same, giving

$$\tilde{\mu}_{ii} = \tilde{\mu}_i \qquad (9.42)$$

where the double subscript denotes the pure phase existing in equilibrium with its corresponding component in the solution.

If the system undergoes an isothermal change, then equation (8.42) holds for the pure phase. Since this equation defines the fugacity of the pure phase and since $\mathrm{d}\tilde{\mu}_{ii} = \mathrm{d}\tilde{\mu}_i$, we may define the fugacity of component i in the solution to be equal to that of the pure component existing in equilibrium with the solution. Thus

$$\mathrm{d}\tilde{\mu}_{i,T} = \tilde{R}T\,\mathrm{d}\,(\ln \tilde{f}_i)_T \qquad (9.43)$$

The limiting condition for $p \to 0$ now becomes

$$\lim_{p \to 0} \frac{\tilde{f}_i}{y_i p} = 1 \qquad (9.44)$$

since as $p \to 0$, $p_i \to y_i p$.

The fugacity of a component in a solution as defined above is not a true partial property. Nevertheless, the notation is retained for partial properties as a reminder that the substance referred to is a component of the solution. However, in the sense that f can be considered as a pseudo-pressure, we may refer to $\tilde{f}_i$ as a pseudo-partial pressure and hence a pseudo-partial property. Now since there is no meaning attached to the expression 'molal fugacity', we shall not use the tilde above the symbol f any more as this would denote that it was a molal quantity.

From equation (8.48), the effect of pressure on the fugacity of component i in a solution at constant temperature is

$$\left(\frac{\partial \ln f_i}{\partial p}\right)_{T,n} = \frac{\tilde{v}}{\tilde{R}T}$$

Now for a pure phase at the same temperature and pressure

$$d\tilde{\mu}'_{iT} = \tilde{R}T d (\ln f'_i)_T \tag{9.45}$$

Differentiating this with respect to p and noting that

$$\left(\frac{\partial \tilde{\mu}'_i}{\partial p}\right)_T = \tilde{v}'_i$$

yields

$$\left(\frac{\partial \ln f'_i}{\partial p}\right)_T = \frac{\tilde{v}'_i}{\tilde{R}T} \tag{9.46}$$

Hence

$$\tilde{R}T \, d \ln \left(\frac{f_i}{f'_i}\right) = (\tilde{v}_i - \tilde{v}'_i) \, dp \tag{9.47}$$

Integrating this from zero pressure to p at constant temperature and composition, yields

$$\tilde{R}T \left[\ln \left(\frac{f_i}{f'_i}\right)\right]^p_{p=0} = \int_0^p (\tilde{v}_i - \tilde{v}'_i) \, dp \tag{9.48}$$

Now as $p \to 0, f'_i \to p$ and $f_i \to y_i p$, so that

$$\tilde{R}T \ln \frac{f_i}{y_i f'_i} = \int_0^p (\tilde{v}_i - \tilde{v}'_i) \, dp \tag{9.49}$$

This equation expresses the relation between the fugacity of component i in a solution and its pure phase when existing at the same temperature and pressure as the solution.

For an ideal solution

$$\tilde{v}_i - \tilde{v}'_i = 0$$

(by definition), therefore

$$(f_i = y_i f'_i)_{p,T} \tag{9.50}$$

This result, which determines the mole fraction of a component in an ideal solution, is known as the Lewis–Randall rule. Notice that at low pressures, this tends to the pressure ratio as shown earlier for ideal gases.

It will now be shown that for all ideal solutions the enthalpy of a component remains unchanged when it goes into solution.

Recall equation (8.51) which gives the effect of temperature on fugacity as

$$\left(\frac{\partial \ln f}{\partial T}\right)_p = - \frac{\tilde{h} - \tilde{h}^\circ}{\tilde{R}T^2}$$

Applying this relation to component i in an ideal solution and to its pure phase existing at the same temperature and pressure, yields

$$\left(\frac{\partial \ln f_i}{\partial T}\right)_{p,n} - \left(\frac{\partial \ln f_i'}{\partial T}\right)_p = \frac{\tilde{h}_i' - \tilde{h}_i}{\tilde{R}T^2} \qquad (9.51)$$

Combining the LHS terms, gives

$$\left[\frac{\partial \ln\left(\frac{f_i}{f_i'}\right)}{\partial T}\right]_p = \frac{\tilde{h}_i' - \tilde{h}_i}{\tilde{R}T^2} \qquad (9.52)$$

By virtue of Lewis–Randall rule, the LHS is zero, zince y_i, the mole fraction of component i in solution, does not depend on temperature, hence

$$\tilde{h}_i - \tilde{h}_i' = 0$$

or

$$\Delta H = \Sigma\, n_i\,(\tilde{h}_i - \tilde{h}_i') = 0 \qquad (9.53)$$

Thus, the earlier remark in Section 9.1 that constancy of volumes and enthalpies are consequences of each other is hereby confirmed. Notice now that the values $\tilde{h}_i'$ represent actual enthalpies of the components and no assumption of ideal gas behaviour has been made.

In Section 9.3 it was shown that for ideal gases the change in entropy on mixing is not zero, and in the following discussion this change will be computed for real gases forming ideal solutions. Beginning with the definition for Gibbs free energy, i.e.

$$\tilde{g} = \tilde{h} - T\tilde{s}$$

and applying this to component i in the solution, in its pure phase at the same pressure and temperature as the solution, this gives

$$\tilde{s}_i - \tilde{s}_i' = \frac{\tilde{h}_i - \tilde{h}_i'}{T} - \frac{\tilde{g}_i - \tilde{g}_i'}{T} \qquad (9.54)$$

But, by equation (8.43)

$$\tilde{g} = \tilde{g}^\circ + \tilde{R}T \ln\frac{f}{f^\circ}$$

Hence

$$\tilde{s}_i - \tilde{s}_i' = \frac{\tilde{h}_i - \tilde{h}_i'}{T} - \tilde{R} \ln\frac{f_i}{f_i'}$$

or

$$\tilde{s}_i - \tilde{s}_i' = \frac{\tilde{h}_i - \tilde{h}_i'}{T} - \tilde{R} \ln\frac{f_i}{y_i f_i'} - \tilde{R} \ln y_i \qquad (9.55)$$

For an ideal solution, the first and second terms on the right are zero, therefore the entropy change becomes

$$\tilde{s}_i - \tilde{s}_i' = - \tilde{R} \ln y_i \qquad (9.56)$$

Hence, the total entropy change on forming the solution is given by

$$\Delta S = \Sigma n_i (\tilde{s}_i - \tilde{s}'_i) = - \tilde{R} \Sigma n_i \ln y_i \qquad (9.57)$$

Note that this is the same result as for ideal gas mixtures, though now the actual entropies and not the ideal gas entropies are to be used.

9.8 EQUILIBRIUM WITH RESPECT TO TRANSFER OF COMPONENTS IN TERMS OF FUGACITY

Consider a system consisting of two phases of a pure substance. Let these two phases be denoted as L and V, say. If these two phases are in equilibrium, then the temperature and pressure of both phases must be equal and, in addition, the chemical potential of both phases are equal.

Integrating the equation

$$d\tilde{\mu}_T = \tilde{R}T \, d (\ln f)_T \qquad (8.42)$$

between the two phases yields

$$\tilde{\mu}^L - \tilde{\mu}^V = \tilde{R}T \, (\ln f^L - \ln f^V) = 0 \qquad (9.58)$$

Hence, an alternative criterion for equilibrium of two phases can be stated as

$$f^L = f^V = f^{\text{sat}} \qquad (9.59)$$

since the equilibrium condition corresponds to the saturated state at a given temperature and pressure.

9.9 EFFECT OF PRESSURE ON VAPOUR PRESSURE

The fugacity of a pure compressed liquid or solid at pressures moderately greater than its saturation pressure can be deduced as follows.

$$d\tilde{g}_T = \tilde{R}T \, d (\ln f)_T = \tilde{v} \, dp_T \qquad (8.42 \, \& \, 8.37)$$

Integrating at constant temperature between the saturated state and that of a higher pressure p, yields

$$\tilde{R}T \int_{f^{\text{sat}}}^{f} d (\ln f) = \int_{p^{\text{sat}}}^{p} \tilde{v} \, dp$$

i.e.

$$RT \ln \frac{f}{f^{\text{sat}}} = \int_{p^{\text{sat}}}^{p} \tilde{v} \, dp \qquad (9.60)$$

Since liquids and solids are nearly incompressible, we can assume $\tilde{v} \approx$ constant.

Hence

$$\tilde{R}T \ln \frac{f}{f^{sat}} \approx \tilde{v}(p - p^{sat}) \qquad (9.61)$$

Since $\tilde{v}$ is small for solids or liquids, the quantity $\tilde{v}(p - p^{sat})$ is small for moderate pressures, and further assuming ideal gas behaviour, this gives

$$\tilde{v}^L (p - p^{sat}) = \tilde{R}T \ln \frac{p^V}{p^{sat}}$$

or

$$\frac{p^V}{p^{sat}} = \exp \left\{ \frac{\tilde{v}^L}{\tilde{R}T} (p - p^{sat}) \right\} \qquad (9.62)$$

For small $p - p^{sat}$, this becomes

$$\frac{p^V}{p^{sat}} = 1 + \frac{\tilde{v}^L}{RT} (p - p^{sat}) \qquad (9.63)$$

Thus, the effect of pressure on the liquid is to increase its vapour pressure. This is known as the Poynting effect.

9.10 RAOULT'S LAW

For a binary solution existing in both the liquid and vapour phases, we have for each component in each phase

$$f_A^V = y_A f'^V_A \quad \text{and} \quad f_B^V = y_B f'^V_B$$

$$f_A^L = x_A f'^L_A \quad \text{and} \quad f_B^L = x_B f'^L_B$$

where x denotes the mole fraction in the liquid phase and y in the gaseous phase. At equilibrium

$$f_A^V = f_A^L \quad \text{and} \quad f_B^V = f_B^L \qquad (9.64)$$

hence

$$y_A f'^V_A = x_A f'^L_A \quad \text{and} \quad y_B f'^V_B = x_B f'^L_B$$

or

$$\frac{y_A}{x_A} = \frac{f'^L_A}{f'^V_A} \quad \text{and} \quad \frac{y_B}{x_B} = \frac{f'^L_B}{f'^V_B} \qquad (9.65)$$

Thus, the mole fractions of the component in the two phases are inversely proportional to the fugacity of the pure component in these two phases at the pressure and temperature of the solution.

If we further assume that the pressure is not too high so that the fugacities can be approximated by the pressures, then

$$f'^{V}_{A} = p$$

and

$$f'^{L}_{A} = f'^{sat}_{A} = p^{sat}_{A}$$

Hence

$$x_{A}p^{sat}_{A} = y_{A}p = p_{A} \tag{9.66}$$

Similarly

$$x_{B}p^{sat}_{B} = y_{B}p = p_{B} \tag{9.67}$$

This rule, which holds for ideal solutions subjected to further restrictions, is known as Raoult's law. It says, in effect, that the partial pressure of a given component in the vapour, p_i, is equal to the product of its mole fraction in the liquid solution, x_i, and its saturation pressure at the given temperature, p_i^{sat}. Thus, according to Raoult's law, when a solute is added to a volatile solvent, the volatility of the solvent is reduced, because its mole fraction is reduced. Moreover, the effect on the volatility is the same for an equal number of moles of any substance which is capable of dissolving in the solvent, as this depends on the mole fractions and not on the characteristics of the solutes. If Raoult's law remains valid for a binary solution over all compositions, then the pressure–composition diagram could be drawn with straight lines.

9.11 HENRY'S LAW

In the present section, it will be shown that Henry's law is a direct consequence of Raoult's law. However, to enable this to be done, we need a thermodynamic relation generally known as the Duhem–Margules relation. We shall therefore first deduce this from the Gibbs–Duhem equation derived in Section 8.3.

Recall that the Gibbs–Duhem equation for a given phase is

$$S dT - V dp + \Sigma n_i d\tilde{\mu}_i = 0 \tag{8.23}$$

For changes between equilibrium states of a binary solution without changes in pressure and temperature, this becomes

$$n_1 d\tilde{\mu}_1 + n_2 d\tilde{\mu}_2 = 0 \tag{9.68}$$

where

$$d\tilde{\mu}_i = d\tilde{\mu}^{V}_i$$

If we assume the vapour of each component to be an ideal gas whose partial pressure in the vapour phase is p^{V}_i, then

$$d\tilde{\mu}^{V}_{i\,p,T} = \tilde{R}T d (\ln p^{V}_i)_{p,T} \tag{9.69}$$

Substituting (9.69) in (9.68) and dividing through by $(n_1 + n_2) \, \check{R}T \mathrm{d}x_1$, yields

$$\frac{n_1}{n_1 + n_2} \frac{\partial \ln p_1^V}{\partial x_1} + \frac{n_2}{n_1 + n_2} \frac{\partial \ln p_2^V}{\partial x_1} = 0$$

Noting the liquid phase is being considered and that $x_1 + x_2 = 1$

$$x_1 \frac{\partial \ln p_1^V}{\partial x_1} - x_2 \frac{\partial \ln p_2^V}{\partial x_2} = 0 \qquad (9.70)$$

or

$$\frac{\partial \ln p_1^V}{\partial \ln x_1} = \frac{\partial \ln p_2^V}{\partial \ln x_2} \qquad (9.71)$$

Both equations (9.70) and (9.71) are known as the Duhem–Margules relation. If Raoult's law is applied to this for the solvent

$$\ln p_1^V = \ln x_1 + \ln p_1^{\mathrm{sat}}$$

whence

$$\frac{\partial \ln p_1^V}{\partial \ln x_1} = 1$$

Hence

$$\frac{\partial \ln p_2^V}{\partial \ln x_2} = 1$$

which gives, on integration

$$p_2^V = k_2 x_2 \qquad (9.72)$$

where k_2 is a constant of integration. This result is known as Henry's law. It applies to the solute.

Notice that Raoult's law gives a similar proportionality relation for the solvent, the constant being the saturation vapour pressure of the pure solvent. However, the proportionality constant k for the solute is not necessarily equal to the saturation vapour pressure of the pure solute.

9.12 LAW OF BOILING POINT ELEVATION

Raoult's law gives a functional relationship among the three variables p_1^{sat}, p_1^V and x_1. Now since p_1^{sat} is uniquely related to the saturation temperature or boiling point of the solvent, T, therefore

$$T = f(p_1^{\mathrm{sat}}) = f'(p_1^V, x_1)$$

From this functional relationship, the rate of change of T with respect to the mole fraction of the solvent can be deduced as follows

$$\left(\frac{\partial T}{\partial x_1} \right)_{p_1^V} = \frac{\mathrm{d}T}{\mathrm{d}p_1^{\mathrm{sat}}} \left(\frac{\partial p_1^{\mathrm{sat}}}{\partial x_1} \right)_{p_1^V} \qquad (9.73)$$

From Raoult's law, if p_1^Y is held constant

$$x_1 \left(\frac{\partial p_1^{sat}}{\partial x_1} \right)_{p_1^Y} + p_1^{sat} = 0$$

Hence, equation (9.73) becomes

$$\left(\frac{\partial T}{\partial x_1} \right)_{p_1^Y} = - \frac{p_1^{sat}}{x_1} \frac{dT}{dp_1^{sat}} \qquad (9.74)$$

For a binary solution, $x_1 + x_2 = 1$, and equation (9.74) becomes

$$\left(\frac{\partial T}{\partial x_2} \right)_{p_1^Y} = \frac{p_1^{sat}}{(1 - x_2)} \frac{dT}{dp_1^{sat}}$$

If the solution is very dilute such that $x_2 \ll 1$ and $p_1^Y \approx p_1^{sat}$, the rate of change of T with respect to x_2 is almost independent of p_1^Y; the above equation can therefore be written as

$$\frac{dT}{dx_2} = p_1^{sat} \frac{dT}{dp_1^{sat}} \qquad (9.75)$$

On substituting for dT/dp_1^{sat} from the Clapeyron equation (equation 3.55), equation (9.75) becomes

$$\frac{dT}{dx_2} = \frac{T p_1^{sat} v_{fg}}{h_{fg}} \qquad (9.76a)$$

If we neglect v_f compared to v_g for the solvent and further assume ideal gas behaviour for the vapour, equation (9.76a) becomes

$$\frac{dT}{dx_2} = \frac{RT^2}{h_{fg}} \qquad (9.76b)$$

Since the quantities on the right hand side of equation (9.76) are fixed at a given external pressure, the above result states, in effect, that the boiling point of a solvent is raised by dissolving in it a nonvolatile solute; and the magnitude of the increase is directly proportional to the amount of solute present.

9.13 OSMOTIC PRESSURE

Osmosis is the name given to the spontaneous flow of solvent into solution when the two are separated by a semi-permeable membrane, which allows the

192

passage of only the solvent molecules. Osmotic pressure is defined as the excess pressure that must be applied to the solution to prevent osmosis.

Consider the system shown in Figure 9.2, where the pure solvent 1 is in equilibrium with both the solution containing a nonvolatile solute 2 as well as the vapour of the pure solvent. Referring to this figure, difference in the chemical potential of 1 between levels S and SS may be expressed in terms of the vapour pressure of 1^V at these levels as

$$(\tilde{\mu}_1^S - \tilde{\mu}_1^{SS})^V = \tilde{R}T \ \ln \ \frac{(p_1^V)^S}{(p_1^V)^{SS}} \tag{9.77}$$

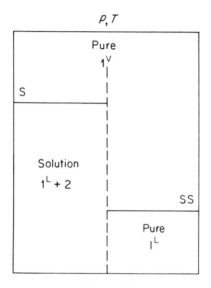

Figure 9.2

Since

$$(p_1^V)^{SS} = p_1^{sat}$$

then by Raoult's law, equation (9.77) becomes

$$(\tilde{\mu}_1^S - \tilde{\mu}_1^{SS})^V = \tilde{R}T \ \ln x_1$$

Also, at constant T, by equation (3.44)

$$d\tilde{\mu}_T = \tilde{v} \, dp_T$$

Integrating this relation over the liquid phases

$$(\tilde{\mu}_1^S - \tilde{\mu}_1^{SS})^L = \tilde{v}_1 \left\{ (p_1^L)^S - (p_1^L)^{SS} \right\}$$

For equilibrium, the chemical potentials of the liquid and vapour phases must be equal. Hence

$$\tilde{R}T \ln x_1 \;=\; \tilde{v}^L \{(p^L)^S - (p^L)^{SS}\} \;=\; -P\tilde{v}^L \tag{9.78}$$

where P denotes the osmotic pressure.

For very dilute solutions

$$\ln x_1 \;=\; \ln(1 - x_2) \;=\; -x_2$$

therefore

$$\tilde{R}Tx_2 \;=\; P\tilde{v}^L$$

or

$$\frac{P}{c_2\,\tilde{R}T} \;=\; 1 \tag{9.79}$$

where $c_2 = x_2/\tilde{v}^L$ represents the concentration of the solute in the solution in moles per unit volume.

9.14 CHEMICAL POTENTIALS OF COMPONENTS IN IDEAL SOLUTIONS

In the last chapter, it was shown that the chemical potential of a gaseous phase can be evaluated by means of an isothermal process. From the foregoing presentation, it is further seen that the idea of fugacity, initially defined to retain the simplicity in form of the expression for chemical potential for real gases, can be extended to solid or liquid phases in equilibrium with the gaseous phase. Thus, the chemical potential of a component in a solution can be expressed in terms of the partial fugacity of the vapour phase in equilibrium with the solution as

$$\tilde{\mu}_1 \;=\; \tilde{R}T \ln f_1^V + F_1(T) \tag{9.80}$$

for the solvent, and

$$\tilde{\mu}_2 \;=\; \tilde{R}T \ln f_2^V + F_2(T) \tag{9.81}$$

for the solute

In view of the numerous further assumptions introduced in earlier discussions, it may be appropriate to introduce some terminologies in parallel to those used with ideal gases. Any solution whose gaseous phase obeys the ideal gas law will be called a perfect solution. With this terminology, ideal gases form perfect solutions and become a subset of our treatment on ideal solutions. In ideal solutions involving coexisting phases, the above definition categorizes a perfect solution as one obeying Raoult's and Henry's laws, the law of boiling point elevation, the law of osmotic pressure and the Poynting effect. Two phase solutions that deviate from perfect solution behaviour will be referred to as semi-perfect solutions. These, as was shown, will obey the Lewis–Randall rule.

From the above discussions it follows that for perfect solutions, equations (9.80) become

$$\tilde{\mu}_1 = \tilde{R}T\ln p_1^{\text{sat}} + \tilde{R}T\ln x_1 + F_1(T) = \tilde{R}T\ln x_1 + \tilde{\mu}_1^\circ \qquad (9.82)$$

and

$$\tilde{\mu}_2 = \tilde{R}T\ln k_2 + \tilde{R}T\ln x_2 + F_2(T) = \tilde{R}T\ln x_2 + \tilde{\mu}_2^\circ \qquad (9.83)$$

where $\tilde{\mu}_1^\circ$ and $\tilde{\mu}_2^\circ$ are functions of temperature only, provided pressure has little or no effect on the saturation vapour pressure of the liquid. Moreover, it is seen that $\tilde{\mu}_1^\circ$ corresponds to the chemical potential of the pure solvent at the pressure and temperature of the solution, and $\tilde{\mu}_2^\circ$ to the chemical potential of the pure solute in a hypothetical dilute solution with $x_2 = 1$.

Generally, it is found from experimental evidence that actual solutions approach perfect behaviour when greatly diluted. This can be inferred from the vapour–pressure diagrams in Example 9.3. On this basis, the further simpli-fication of equation (9.78) using the dilute solution assumption is therefore justified. One should note that most, if not all, of the experimentally observed laws in physical chemistry hold for perfect solutions only. The Lewis–Randall rule, besides our basic definition for ideal solutions, is the only relation that governs the behaviour of semi-perfect solutions.

9.15   NON-IDEAL SOLUTIONS

It is fundamental, according to our definition of ideal solutions, that the change in volume on the formation of a non-ideal solution, given by

$$\Delta V = V_{\text{mixture}} - \Sigma V_{\text{components}}$$

is not zero. A direct consequence of this is that the change in enthalpy on mixing is also not zero. As to the change in entropy, it is now given by equation (9.55) in full.

In addition, we have seen in Section 9.7 that our definition of an ideal solution leads directly to a simple relation between the fugacities of a component in the solution and its pure phase at the same pressure and temperature. This relation is known as the Lewis–Randall rule. Thus, as an approach to handling non-ideal solutions, a non-ideal solution may be alternatively defined as one for which the Lewis–Randall rule does not hold for at least one of the components in the solution. Of course, other relations deduced solely on thermodynamic grounds without any simplifying assumptions should remain valid. These include all the theories presented in the last Chapter. Also, the Duhem–Margules relation as given by equations (9.70) and (9.71) will be valid if the vapour pressure of the components is low. However, this relation will be precisely obeyed if we used partial fugacities instead of the partial pressures in these equations.

Following this latter definition, and prompted by its simplicity, a correction factor is introduced such that the Lewis–Randall rule may continue to apply to a non-ideal solution in the form

$$f_i = \gamma_i x_i f_i' \qquad (9.84)$$

This correction factor is called the activity coefficient of component i in the solution. It depends, in general, on the temperature, pressure and composition of the solution. Obviously, for an ideal solution $\gamma_i = 1$.

Conforming to general usage, the activity of component i will be introduced via the definition

$$a_i = \gamma_i x_i \qquad (9.85)$$

or

$$a_i = \frac{f_i}{f_i'} \qquad (9.86)$$

Thus, the activity of a component can be regarded as the ratio of the fugacity of the component in solution to its fugacity in the pure state at the same temperature and pressure. Note that both the activity and activity coefficient of a component are dimensionless quantities.

In terms of the activity, the chemical potential of component i can be written as

$$\tilde{\mu}_i = \tilde{R} T \ln f_i^V + F_1(T)$$

$$= \tilde{R} T \ln \gamma_i x_i f_i'^V + F_1(T)$$

$$= \tilde{R} T \ln a_i + \tilde{R} T \ln f_i'^V + F_1(T)$$

$$= \tilde{R} T \ln a_i + \tilde{\mu}_i' \qquad (9.87)$$

where $\tilde{\mu}_i'$ represents the chemical potential of the pure component i at the temperature and pressure of the solution.

Since in any state the activity at a given temperature is always proportional to the fugacity, it is easy to show that the effect of pressure on activity is identical to that of pressure on fugacity. Likewise, at any temperature for a given pressure the same relation exists between activity and fugacity and hence the effect of temperature on activity is also identical to that of temperature on fugacity. Thus equations (8.48) and (8.51) apply to activity as well. When the composition is held constant, $d(\ln a_i) = d(\ln \gamma_i)$; hence exactly similar equations hold for the activity coefficient also.

Before concluding this brief introduction to non-ideal solutions, students are referred back to Section 9.5 for an empirical treatment of real gases forming non-ideal solutions.

EXAMPLES

Example 9.1 Show that for an ideal solution the volume–composition and enthalpy–composition diagrams are straight lines.

From our definition of an ideal solution

$$V = \Sigma V' = \Sigma m_i v_i$$

Dividing by the total mass of the system yields

$$\frac{V}{m} = v = \Sigma \frac{m_i}{m} v_i = \Sigma x_i v_i$$

where x_i is the mass fraction of component i. Thus, the v versus x diagram is a straight line. A similar relation can be derived for h.

It was shown in Example 8.1 that the partial property of a component in a binary solution can be obtained from the intercept made by the tangent at the required composition on the pure component ordinate. It is therefore seen that for an ideal solution the partial volume and enthalpy are independent of the composition and equal the specific property of the component.

Example 9.2 A mixture consisting of 75 percent hydrogen and 25 percent nitrogen by volume at 1 atm and 283 K is compressed from an initial volume of 7.07 m³ to a final volume of 28.32 litres. The pressure as measured at a temperature of 323 K is 348 atm. Check this experimental result against values computed from (a) an ideal gas mixture, (b) an ideal solution obeying the law of additive pressures, and (c) an ideal solution obeying the law of additive volumes; for (b) and (c) use van der Waals' equation of state and the generalized compressibility chart to describe the behaviour of the individual gases.

(a) Since an ideal gas mixture behaves as an ideal gas and the number of moles of each gas remains constant throughout

$$p_2 = p_1 \frac{V_1 T_2}{V_2 T_1} = 1 \times \frac{7.07}{0.02832} \times \frac{323}{283} = 285 \text{ atm}$$

(b) Using van der Waals equation, by equation (9.28)

$$p = \tilde{R}T \left(\Sigma \frac{y_i}{\tilde{v} - y_i b_i} \right) - \frac{1}{\tilde{v}^2} \Sigma a_i y_i^2$$

Since

$$N = \frac{pV}{\tilde{R}T} = \frac{101.325 \times 7.07}{8.3143 \times 283} = 0.304 \text{ kmol}$$

$$\tilde{v} = \frac{28.32}{304} = 0.0931 \, l/\text{mol}$$

therefore

$$p = \frac{8.3143}{101.325} \times 323 \left(\frac{0.75}{0.0931 - 0.75 \times 0.0265} + \frac{0.25}{0.0931 - 0.25 \times 0.0385} \right)$$

$$-\frac{1}{0.0931^2}(0.244 \times 0.75^2 + 1.347 \times 0.25^2)$$

$$= 351 - 25.5$$

$$= 325.5 \text{ atm}$$

(c) Using van der Waals' equation, by equation (9.2)

$$0.75 \, \tilde{v}_H + 0.25 \, \tilde{v}_N = 0.0931$$

or

$$\tilde{v}_H = 0.1241 - 0.3333 \, \tilde{v}_N$$

The solution to the problem consists in taking trial values of $\tilde{v}_N$, obtaining $\tilde{v}_H$ from above expression and then calculating p for each gas using van der Waals' equation. The criterion of correct trial values is equality of the pressures computed for the two gases.

For example, assuming $\tilde{v}_N = 0.094 \; l/mol$ gives $\tilde{v}_H = 0.0928 \; l/mol$.

The pressures calculated from van der Waals' equation are

$$p_N = 328 \text{ atm}$$

and

$$p_H = 372 \text{ atm}$$

Thus p_N must be increased and p_H decreased. An examination of van der Waals' equation reveals that this can be brought about by decreasing $\tilde{v}_N$ and increasing $\tilde{v}_H$. Carrying out this procedure a few times, the final values for $\tilde{v}_N$ and $\tilde{v}_H$ are found to be

$$\tilde{v}_N = 0.0881 \; l/mol$$

and

$$\tilde{v}_H = 0.0947 \; l/mol$$

giving

$$p_N = 363 \text{ atm}$$

and

$$p_H = 364 \text{ atm}$$

As these values are close enough, the average value for p may be taken as 363.5 atm.

(b) with the generalized z–chart

Assuming $p = 350$ atm, the reduced temperature and pressure for each component are
for hydrogen

$$T_r = \frac{323}{33.2} = 9.73$$

$$p_r = \frac{350}{12.8} = 27.3$$

giving

$$z_H = 1.30$$

for nitrogen

$$T_r = \frac{323}{126} = 2.56$$

$$p_r = \frac{350}{33.5} = 10.44$$

giving

$$z_N = 1.18$$

therefore

$$z_{mix} = 0.75 \times 1.30 + 0.25 \times 1.18 = 1.270$$

whence

$$p = \frac{1.27 \times 8.3143 \times 323}{0.0931 \times 101.325} = 361 \text{ atm}$$

The iteration process may now be continued until the assumed and computed pressure values are close enough.
Next try

$$p = 361 \text{ atm}$$

for hydrogen

$$p_r = \frac{361}{12.8} = 28.2$$

$$z_H = 1.31$$

for nitrogen

$$p_r = \frac{361}{33.5} = 10.78$$

$$z_N = 1.17$$

therefore

$$z_{mix} = 0.75 \times 1.31 + 0.25 \times 1.17 = 1.274$$

giving

$$p = 361 \times \frac{1.274}{1.27} = 362 \text{ atm.}$$

(c) with the generalized z–chart, since

$$z_i = \frac{p_i \tilde{v}_i}{\bar{R}T}$$

and

$$\tilde{v}_i = \frac{\tilde{v}_m}{x_i}$$

therefore

$$z_i = \frac{p_i \tilde{v}_m}{x_i \bar{R} T}$$

Thus

$$z_H = \frac{0.0931 \quad p_H}{0.75 \times 8.3143 \times 323}$$

which gives on rearrangement and simplification

$$p_H = 213.5 \ z_H$$

If a generalized z-chart in v_r and T_r is available, the procedure is straight-forward, z_H may be found from its reduced volume and temperature values, then substituted into above expression to get p_H; p_N may be found in a similar manner. The total pressure is then given by $p_H + p_N$.

If a generalized z-chart in p_r and T_r only is available, one must iterate on z by guessing a value for p_H. Assume $p_H = 250$ atm, say, then

$$T_r = 9.73$$

$$p_r = \frac{250}{12.8} = 19.53$$

giving

$$z_H = 1.21$$

(from the generalized z-chart). But z_H computed from the assumed value of p_H is

$$z_H = \frac{250}{213.5} = 1.17$$

On the next try, therefore one must increase the computed value of z_H. This obviously is obtained by increasing the assumed p_H. As the next guess, in fact, the value of p_H may be taken from the expression derived above, by substituting the value found from the generalized z-chart in the last trial into this relation. By so doing, the value of p_H for the next trial is $213.5 \times 1.21 = 258$ atm.

Using this value of p_H,

$$p_r = \frac{258}{12.9} = 20.0$$

the new z_H value from the generalized z-chart is approximately 1.21. Similarly, we find that

$$p_N = 71.1 \ z_N$$

Assume $p_N = 100$ atm

$$z_N \text{ (computed)} = \frac{100}{71.1} = 1.406$$

But

$$T_r = 2.56$$

$$p_r = \frac{100}{33.5} = 2.98$$

giving

$$z_N = 1.025$$

from the generalized z-chart.

For next trial, use

$$p_N = 71.1 \times 1.025 = 72.9 \text{ atm}$$

giving

$$p_r = \frac{72.9}{33.5} = 2.175$$

whence

$$z_N = 0.98.$$

For the third trial, take $p_N = 71.1 \times 0.98 = 69.6$ atm
giving

$$p_r = \frac{69.6}{33.5} = 2.08$$

whence

$$z_N = 0.98 \text{ (approximately)}.$$

Hence

$$p = 258 + 69.6 = 327.6 \text{ atm}$$

Discussion   A comparison of the above results with the experimentally obtained value shows that the ideal gas assumption gives the worst result. The degree of accuracy between the generalized compressibility chart and van der Waals' equation is about the same. For higher accuracies, one must resort to equations of state making use of more empirical constants. This will make the computation work much more tedious and burdensome. Noting that both nitrogen and hydrogen have z_c values higher than that for the generalized z-chart, the use of the z-chart for Group III gases will also increase the accuracy.

Example 9.3   Discuss the features of the vapour pressure–composition diagram of a typical binary solution and identify the regions in which the various thermo-dynamic 'laws' derived in this chapter hold.

A typical vapour pressure–compsition diagram of a binary solution of two completely miscible substances is shown in Figure 9.3. The dashed lines on the diagram indicate perfect solution behaviour over the entire range of the composition. However, the actual solution departs from this over a large inter-mediate range. The letters R and H by the sides of the diagram indicate Raoult's law and Henry's law behaviour of the solvent and solute respectively. It is

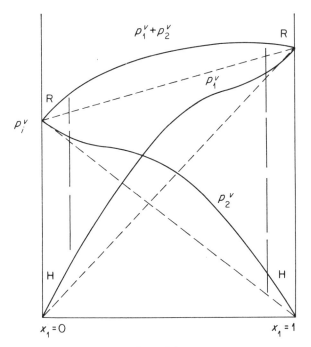

Figure 9.3

therefore seen that these laws hold only for dilute or perfect solutions.

However, using the Duhem–Margules relation, it can be readily shown that the qualitative characteristics of the curves in the intermediate region where neither Raoult's nor Henry's law holds are consistent with each other. Thus, the Duhem–Margules relation is of more general application than Raoult's or Henry's law.

Example 9.4 It is known that at 298 K the density of ethanol is 785.1 kg/m^3 and that of water is 997.0 kg/m^3. Also, the partial molal volumes of ethanol and water are respectively 57.30×10^{-6} and 17.15×10^{-6} m^3/mol in an ethanol–water mixture which is 60 percent ethanol by weight. Determine the volume of 1 kg of this mixture assuming that (a) the mixture is ideal, and (b) the mixture is not ideal. The molecular weights of ethanol and water are 46.07 and 18.02 respectively.

(a) For an ideal mixture

$$V = \Sigma V' = \frac{0.6}{785.1} + \frac{0.4}{997.0} = 1.1654 \times 10^{-3} \text{ m}^3$$

(b) In one kg of this mixture, the number of moles of ethanol

$$= \frac{600}{46.07} = 13.02 \text{ mol}$$

and the number of moles of water

$$= \frac{400}{18.02} = 22.20 \text{ mol}$$

Therefore
$$V = \Sigma n_i \tilde{v}_i = 13.02 \times 57.30 \times 10^{-6} + 22.20 \times 17.15 \times 10^{-6}$$

$$= 1.1268 \times 10^{-3} \text{ m}^3$$

Hence, the contraction in volume on mixing is

$$\frac{1.1654 - 1.1268}{1.1268} = 3.43 \text{ percent}$$

of the actual volume at this given concentration.

PROBLEMS

1. The boiling point of a solute is raised by 0.1 K when 1 g of A is dissolved in 1 kg of water at a temperature of 287 K. What is the molecular weight of A in water solution?

2. A closed vessel contains a mixture of air and water vapour in contact with an excess of water. The pressures in the vessel at 300 K and 333 K are respectively 777 mm and 981 mm of mercury. If the vapour pressure of water at 300 K is 27 mm of mercury, what is the vapour pressure at 333 K?

3. Find the mass of water vapour per m^3 of moist air when the temperature is 293 K and the dew-point 283 K. Use the following data: density of air at 273 K and 760 mm mercury pressure is 1.29 g/l, density of water vapour is 5/8 that of air at the same temperature and pressure; saturation vapour pressure of water vapour at 283 K is 9.20 mm mercury.

4. A vessel of volume $2V$ is divided into two equal compartments. These are filled with the same perfect gas, the temperature and pressure on one side of the partition being (p_1, T_1) and on the other (p_2, T_2). Show that if the gases on the two sides are allowed to mix slowly with no heat entering, the final pressure and temperature will be given by

$$p = \tfrac{1}{2}(p_1 + p_2)$$

and

$$T = \frac{T_1 T_2 (p_1 + p_2)}{(p_1 T_2 + p_2 T_1)}$$

Further, show that the entropy gain is

$$V\left[\left(\frac{c_p}{R}\right)\left\{\frac{p_1}{T_1}\ln\frac{T}{T_1} + \frac{p_2}{T_2}\ln\frac{T}{T_2}\right\} - \frac{p_1}{T_1}\ln\frac{p}{p_1} - \frac{p_2}{T_2}\ln\frac{p}{p_2}\right]$$

5. An insulated box is divided into three compartments not necessarily equal in volume, and separated from each other by adiabatic walls. Each compartment contains a perfect gas at pressure and temperature different from the other two. The walls are pierced and the gases allowed to mix. Show that final temperature of the mixture is given by

$$T_m = \frac{m_A c_{vA} T_A + m_B c_{vB} T_B + m_C c_{vC} T_C}{m_A c_{vA} + m_B c_{vB} + m_C c_{vC}}$$

What are the pressure and entropy gain of the mixture?

6. In a steady flow system, two perfect gases initially at temperatures of T_A and T_B are to be mixed adiabatically to form a mixture. Each gas enters the mixing chamber at 1 atm pressure, and the mixing occurs under a constant total pressure of 1 atm. Assuming that kinetic and potential energy changes are negligible, determine (a) the temperature of the mixture, and (b) the irreversibility, if the mixture contains 50 percent by volume of each gas.

7. Show that Raoult's law is valid for component A of a binary solution over the range of composition for which Henry's law holds for B, i.e. show that

$$p_A = p_A^{sat} x_A$$

when

$$p_B = kx_B.$$

8. Show that the freezing point of a liquid is reduced by a solute, the amount of reduction given by

$$\frac{\partial T}{\partial x_1} = -\frac{\tilde{R}T^2}{\tilde{h}_{sf}}$$

where x_1 denotes the mole fraction of the solute and $\tilde{h}_{sf}$ the latent heat of melting per mole of the substance.

Hence show that for a normal solution the freezing point of water is reduced by 1.85 K.

9. One mole of each of the gases A and B and two moles of C are enclosed in a flexible container of volume 6.25 l at a temperature of 311 K. No chemical reactions take place inside the container. Find the total pressure. The container is now lowered to the bottom of the ocean where it collapses to a volume of 0.125 l. How deep is the ocean? The temperature of the ocean at this depth is measured at 283 K.

The properties of the gases are

Gas	Critical pressure (atm)	Critical temperature (K)
A	34	11.1
B	44.2	22.2
C	40.8	111.0

10. Show that if a solvent and its vapour have different molecular weights M_o and M_o' the change in vapour pressure brought about by $n_1, n_2, \ldots, n_n$ mols of n substances dissolved in n_o mols of the solvent is given by

$$\frac{\Delta p}{p} = - \frac{M_o}{M_o'} \frac{\sum\limits_{i=1}^{n} n_i}{n_o}$$

CHAPTER TEN

Thermodynamics of Reacting Systems

10.1 FIRST LAW ANALYSIS OF CHEMICAL REACTIONS

The first law may be applied to a chemical reaction in which $a\Delta\xi$ moles of A combines with $b\Delta\xi$ moles of B to form $k\Delta\xi$ moles of K and $l\Delta\xi$ moles of L. The stoichiometric equation for this reaction can be written as

$$aA + bB \rightleftharpoons kK + lL \qquad (10.1)$$

Since chemical reactions usually occur at constant pressure, this may be visualized as taking place in the form of a steady flow process. However, there are now multiple inlet and outlet streams leading to and from the control volume, two in each case for our present reaction. In order to assure that these streams conduct only a pure component, we may assume that semi-permeable membranes allowing only the relevant component to pass through are instituted at the junction of each stream with the control volume where the reaction occurs.

Applying the steady flow energy equation to the control volume

$$Q = (k\tilde{h}_K + l\tilde{h}_L - a\tilde{h}_A - b\tilde{h}_B)\,\Delta\xi \qquad (10.2)$$

Equation (10.2) permits us to find the heat of reaction in terms of the enthalpies of the pure species. However, tables of properties available do not immediately fulfil the requirements of this equation because the state of zero enthalpy (or energy) for each species is usually arbitrarily set and they may not coincide for all the species involved in the reaction. This difficulty can be removed by referring the enthalpy of each substance to a certain standard state, denoted by the superscript 0. With this innovation, equation (10.2) becomes

$$\frac{Q}{\Delta\xi} = k(\tilde{h}_K - \tilde{h}_K^0) + l(\tilde{h}_L - \tilde{h}_L^0) - a(\tilde{h}_A - \tilde{h}_A^0) - b(\tilde{h}_B - \tilde{h}_B^0)$$
$$+ (k\tilde{h}_K^0 + l\tilde{h}_L^0 - a\tilde{h}_A^0 - b\tilde{h}_B^0) \qquad (10.3)$$

Now, the quantities of the form $(\tilde{h} - \tilde{h}^0)$ are independent of the arbitrary selection of the zero state for enthalpy since this represents a change in the enthalpy values. The quantity $\tilde{h}^0$ in the last term is known as the enthalpy of formation of the substance at the standard state. Values for all $\tilde{h}^0$s must be based on some consistent selection of zero states which can no longer be arbitrary for each of

206

the species. The reason for this can be seen from the following consideration: suppose we assign zero to the enthalpy of formation for the reactants A and B. As the heat of reaction is a unique characteristic of a given reaction, the state of zero $\tilde{h}^\circ$ for the products are also fixed and we are not free to manipulate their values. Conventionally, the standard state for reporting $\tilde{h}^\circ$ values has been chosen at 1 atmosphere (101.325 N/m^2) pressure and a temperature of 298.16 K.

For convenience, the enthalpy of all the elements in their most stable form is assigned the value of zero at the standard state. The enthalpy of formation of a compound is then given by the enthalpy of the compound at the standard state. This is also known as the standardized enthalpy, meaning simply that it is properly related to the enthalpy of other elements and compounds. Values for standardized enthalpies are often given in tables of chemical data. However, it should be noted that the term 'heat of formation' used in many common tables refers sometimes to the enthalpy of formation and sometimes to the negative of the enthalpy of formation. This confusion arises because of the convention adopted in chemistry that heat flowing from a reacting system is positive. The user of any *heat of formation* table should therefore first find out which convention is being used.

The full equipment is now available to apply the first law to chemical reactions. The quantity in the last term of equation (10.2) is called the enthalpy of reaction at the standard state for the stoichiometric reaction. It represents the increase in enthalpy when a system changes from a standard state corresponding to a moles of A, and b moles of B to a standard state corresponding to k moles of K and l moles of L.

Hess law

A direct consequence of the application of thermodynamics to chemical reactions is the reason behind a fundamental experimental law in thermochemistry known as the Hess law. This states that the heat change accompanying a chemical reaction is the same whether the reaction takes place in one or more stages.

The thermodynamic reason behind this law is trivial. It follows immediately if we realize that this heat of reaction is a manifestation of the property change of the system when it goes from one state to another.

The Hess law is useful in calculating enthalpies of reactions that have not been experimentally measured.

10.2 EFFECTS OF CHEMICAL REACTIONS ON NUMBER OF COMPONENTS: CHEMICAL EQUILIBRIUM

Earlier in Section 8.1 the relation between the number of reference properties required to define a system was derived in the form of the Gibbs phase rule, and it was noted that this rule holds for a non-reacting system. The effect an allowed chemical reaction has on the number of components present in the system was briefly mentioned, and a more detail discussion on this topic will now be given.

The problem will first be considered qualitatively. Consider a gaseous phase consisting of n_O moles of O_2, n_H moles of H_2 and n_W moles of H_2O. If the phase is at room temperature, we should not observe any change in the number of moles of the different constituents over a period of days. On this time scale, the phase could be considered in equilibrium. The number of components would be three, since by introducing O_2, H_2 or H_2O, the number of moles of each of these substances present could be altered independently.

If, on the other hand, this phase were observed over a period of years, it would be found that H_2O was forming at the expense of the other two components. On this time scale, one would conclude that the phase is not in equilibrium. The change toward equilibrium could be speeded up by the introduction of a catalyst, such as spongy platinum. When equilibrium is reached, one of the species of O_2 or H_2 would have virtually disappeared, leaving only H_2O and the remainder of these two species.

If the temperature of the phase were raised by one or two thousand kelvins, at equilibrium a large fraction of H_2O and a smaller but considerable amount of both O_2 and H_2 would be found. Under such circumstances the introduction of one mole of H_2O into the phase is entirely equivalent, as regards equilibrium composition at the given temperature, to the addition of one mole of H_2 and one half mole of O_2. Any allowed change in composition could be accomplished by adding or subtracting appropriate amounts of H_2 and O_2. The number of components of the phase is thus two, whereas if chemical reaction had been prevented, the number would be three.

The above observation may be generalized to include more than one allowed reaction among the constituents. Thus, the number of components of a phase when j different chemical reactions may occur between its constituents is j less than the number when all chemical reactions are prevented.

This result can be put on a quantitative basis by considering the chemical equilibrium among the reactants and products of reaction in the system. The stoichiometric equation for the above example is

$$H_2 + \tfrac{1}{2}O_2 \rightarrow H_2O \qquad (10.4)$$

For our present purpose, consider an allowed variation consisting of the formation of dn moles of H_2O from dn moles of H_2 and $\tfrac{1}{2}dn$ moles of O_2. We shall hold the pressure and temperature of the system constant during this variation. Thus

$$dG \geq 0$$

Written in terms of the chemical potentials of each components, this becomes

$$-\tilde{\mu}_H\,dn - \tfrac{1}{2}\tilde{\mu}_O\,dn + \tilde{\mu}_W\,dn \geq 0$$

Now since dn may assume positive or negative values, as the allowed variation may be reversed by decomposing the H_2O to form H_2 and O_2, the inequality cannot hold, and for equilibrium

$$\tilde{\mu}_W = \tilde{\mu}_H + \tfrac{1}{2}\tilde{\mu}_O \qquad (10.5)$$

Thus, it is seen that for each chemical reaction that is possible among the components an *equation of chemical equilibrium* such as (10.5) must exist. Hence, not all the components in the system are independent. The number of independent components will be reduced by one for each possible reaction present.

10.3 EQUILIBRIUM CONSTANT AND THE LAW OF MASS ACTION

Consider the reaction given by (10.4), assuming this to take place in a reaction chamber of the form shown in Fugure 10.1. If each pure phase in equilibrium with the reacting mixture behaves as an ideal gas, then

$$\tilde{\mu}_i = \tilde{R} T \ln p_i + F_i(T)$$

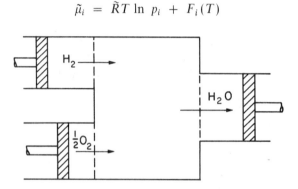

Figure 10.1

On substituting in equation (10.5)

$$\tilde{R} T \ln p_H + F_H(T) + \tfrac{1}{2}\{\tilde{R} T \ln p_O + F_O(T)\} = \tilde{R} T \ln p_{H_2O} + F_{H_2O}(T) \quad (10.6)$$

Hence

$$\tilde{R} T \ln \frac{p_{H_2O}}{p_H \times p_O^{1/2}} = F_H(T) + \tfrac{1}{2} F_O(T) - F_{H_2O}(T) = F_p(T)$$

i.e.

$$\frac{p_{H_2O}}{p_H \times p_O^{1/2}} = \exp\left\{\frac{F_p(T)}{\tilde{R} T}\right\} = K_p(T) \quad (10.7)$$

where $K_p(T)$ is a function of temperature only. At any one value of temperature, it is called the equilibrium constant in terms of partial pressures.

Assuming the gases form an ideal gas mixture in the reaction chamber, then

$$p_H + p_O + p_{H_2O} = p \quad (10.8)$$

and

$$\frac{p_O}{n_O} = \frac{p_H}{n_H} = \frac{p_{H_2O}}{n_{H_2O}} \quad (10.9 \ \& \ 10.10)$$

since each of these equals $\tilde{R}T/v$

Also, for purely chemical reactions, the atoms are indestructible, whence, the balance for the number of atoms present before and after the reaction gives

$$2n_{O,i} + n_{H_2O,i} = 2n_{O,f} + n_{H_2O,f} \tag{10.11}$$

and

$$2n_{H,i} + 2n_{H_2O,i} = 2n_{H,f} + 2n_{H_2O,f} \tag{10.12}$$

Thus, we have a total of six equations, (10.7) to (10.12). These will completely determine the six unknowns: three ps and three n_fs. The equilibrium composition of our system is therefore completely specified.

The result expressed in equation (10.7) can be generalized to a reaction of the form

$$aA + bB + \ldots \rightleftharpoons kK + lL + \ldots$$

whence we have for equilibrium

$$a\tilde{\mu}_A + b\tilde{\mu}_B + \ldots = k\tilde{\mu}_K + l\tilde{\mu}_L + \ldots$$

which gives

$$K_p(T) = \frac{p_K^k \times p_L^l \ldots}{p_A^a \times p_B^b \ldots} \tag{10.13}$$

This result can be put in a form involving the mole fractions of the gases if we assume ideal gas mixture behaviour in the reaction chamber. For an ideal gas mixture

$$y_i = \frac{n_i}{n} = \frac{p_i}{p}$$

hence

$$\frac{y_K^k \times y_L^l \ldots}{y_A^a \times y_B^b \ldots} = K_p(T) \times p^{(a+b+\ldots) - (k+l+\ldots)}$$

$$= K_p(T) \times p^{-\Delta n}$$

$$= K_y(p, T) \tag{10.13a}$$

where Δn is the change in the number of moles in the stoichiometric equation. This relation is generally known as the *law of mass action*. It predicts the effects of temperature and pressure on the equilibrium composition. It is noted from this that if $\Delta n = 0$, there is no change in the number of moles and no change in the volume during the reaction; consequently, pressure has no effect on the equilibrium composition.

Real gas behaviour for the pure phases

If the pressure of the pure phases in equilibrium with the mixture phase is too high or the temperature too low, the chemical potential has to be expressed in terms of the fugacity of the phase. Thus, the symbol f replaces p in equation (10.13) and we get an equilibrium constant in terms of fugacities as

$$K_f(T)\, \frac{f_K^k \times f_L^l \cdots}{f_A^a \times f_B^b \cdots} \tag{10.14}$$

Reaction in the liquid phase

If instead, the reaction proceeds in the liquid phase and the solution can be assumed to be perfect, then

$$\tilde{\mu}_i \; = \; \tilde{R}T \ln x_i + \tilde{\mu}_i'$$

On substituting this into the equation for chemical equilibrium, we get

$$K_x(T)\, \frac{x_K^k \times x_L^l \cdots}{x_A^a \times x_B^b \cdots} \tag{10.15}$$

where K_x is a function of temperature only, provided we can neglect the Poynting effect. The equilibrium constant for ideal solutions which do not obey Raoult's law will be the same as that given by equation (10.14).

For non-ideal solutions, it can be readily shown that the activities replace the mole fractions in equation (10.15). In general, the equilibrium constant is now both a function of pressure and temperature.

Heterogeneous reactions

The equilibrium constant for non-ideal solutions can be applied directly to reactions involving substances in heterogeneous phases also. If gaseous substances are also involved in the reaction, then simplification of the RHS expression can be achieved by assuming the activities of the solid and liquid phases to be unity. This is a valid assumption since the vapour pressures of the liquid and solid phases does not vary much with increase in total pressure; whence $f_i = f_i'$ approximately.

10.4 SYSTEMS IN STABLE CHEMICAL EQUILIBRIUM: LE CHATELIER'S PRINCIPLE

Consider a system in stable chemical equilibrium and let the stoichiometric equation for the reaction be

$$a\mathrm{A} + b\mathrm{B} + \ldots \rightleftharpoons k\mathrm{K} + l\mathrm{L} + \ldots$$

Let

$$(k + l + \ldots) > (a + b + \ldots)$$

so that pressure has an effect on the equilibrium constant of the system at the given temperature. If the pressure of the system in equilibrium is increased and the hypothesis is made that this causes the system to be displaced in a direction towards the formation of more products, then the pressure of the system should be further increased since $\Sigma v_{\text{products}} > \Sigma v_{\text{reactants}}$; hence by the hypothesis, the reaction should proceed further and in the end all the reactants will be converted into products. Thus the system cannot be in stable equilibrium since the pressure increase introduced in the beginning need only be infinitesimally small. This contradicts the hypothesis and hence the effect of the increase in pressure can only be such as to reduce the formation of more products, i.e. to reduce the pressure of the system. Thus we have deduced *Le Chatelier's principle* for the role of pressure.

Similar arguments are applicable to effects of temperature and concentration. The net result of all these can be summarized as follows.

If a change is made in the pressure, temperature or concentration of a system in stable chemical equilibrium, the equilibrium will be displaced in such a direction as to oppose the effect of this change.

10.5 CHEMICAL POTENTIAL AND EQUILIBRIUM CONSTANT OF FORMATION

Consider the formation of a substance B, say, from its elements. The reaction involved can be written as

$$v_1 A_1 + v_2 A_2 + \ldots + v_n A_n \rightleftharpoons B$$

where $v_1, \ldots, v_n$ denote the number of moles of elements $A_1, \ldots, A_n$ which combine to form one mole of B. The condition for chemical equilibrium is, by equation (10.5)

$$\tilde{\mu}_B = v_1 \tilde{\mu}_{A_1} + \ldots + v_n \tilde{\mu}_{A_n}$$

whence

$$\tilde{R} T \ln \alpha_B + \tilde{\mu}_B^o = v_1 \{\tilde{R} T \ln \alpha_1 + \tilde{\mu}_{A_1}^o\} + \ldots + v_n \{\tilde{R} T \ln \alpha_n + \tilde{\mu}_{A_n}^o\}$$

where α denotes p, f, x or a as the case requires. Thus

$$\tilde{R} T \ln \frac{\alpha_B}{\alpha_1^{v_1} \ldots \alpha_n^{v_n}} = (v_1 \tilde{\mu}_{A_1}^o + \ldots + v_n \tilde{\mu}_{A_n}^o - \tilde{\mu}_B^o)$$

The usual convention is to assign the chemical potential of each element as zero when its temperature is that under consideration and its pressure is 1 atmosphere (101.325 kN/m^2) and when it is in its most stable condition at this T and p. Thus, all terms on the RHS of the above equation except $\tilde{\mu}_B^o$ are zero. Furthermore, noting that

$$\frac{\alpha_B}{\alpha_1^{v_1} \ldots \alpha_n^{v_n}} = K_{\text{formation}}$$

we have

$$\tilde{\mu}_B^o = - RT \ln K_{formation} \tag{10.16}$$

Equation (10.16) gives a simple relation between the equilibrium constant of formation of a substance and the chemical potential of this substance in its standard state.

10.6 FREE ENERGY CHANGE AND EQUILIBRIUM CONSTANT

The considerations of the last section can be generalized to any arbitrary stoichiometric reaction of the type

$$aA + bB + ... \rightleftharpoons kK + lL + ...$$

For any allowed variation specified by this reaction at constant p and T

$$dG = (k\tilde{\mu}_K + l\tilde{\mu}_L + ... - a\tilde{\mu}_A - b\tilde{\mu}_B - ...)\,d\xi$$

i.e.

$$\frac{dG}{d\xi} = k(\tilde{\mu}_K - \tilde{\mu}_K^o) + l(\tilde{\mu}_L - \tilde{\mu}_L^o) + ... - a(\tilde{\mu}_A - \tilde{\mu}_A^o) - b(\tilde{\mu}_B - \tilde{\mu}_B^o)$$

$$- ... + k\tilde{\mu}_K^o + l\tilde{\mu}_L^o + ... - a\tilde{\mu}_A^o - b\tilde{\mu}_B^o - ...$$

On substituting

$$\tilde{\mu}_i = \tilde{R}T \ln \alpha_i + \tilde{\mu}_i^o$$

where α denotes p, f, x or a as the case requires, this yields

$$\frac{dG}{d\xi} = \tilde{R}T \ln \frac{\alpha_K^k \times \alpha_L^l \,...}{\alpha_A^a \times \alpha_B^b \,...} + k\tilde{\mu}_K^o + l\tilde{\mu}_L^o + ... - a\tilde{\mu}_A^o - b\tilde{\mu}_B^o - ... \tag{10.17}$$

$$= \tilde{R}T \ln Q + \Delta G^o$$

where ΔG^o represents the free energy change of the stoichiometric reaction at standard state, and Q the argument of the logarithmic term.

The condition for equilibrium requires that

$$\frac{dG}{d\xi} = 0$$

Now the argument of the logarithm is just the equilibrium constant and is denoted by K_α. Therefore

$$\Delta G^o = - \tilde{R}T \ln K_\alpha \tag{10.18}$$

This very important equation in chemical thermodynamics is known as the *van't Hoff isotherm*

For any non-equilibrium composition of the reactive mixture, equation (10.17) holds, since this merely gives the rate of change of the Gibbs free energy of the mixture if the reaction is allowed to proceed forward. Thus the free energy change of an initially non-equilibrium mixture before reaching chemical equilibrium is given by

$$\frac{dG}{d\xi} = \tilde{R}T \ln \frac{Q}{K_\alpha} \tag{10.19}$$

This equation is known as the *reaction isotherm*. The parameter Q is commonly referred to as the *reaction quotient*, and it becomes equal to the equilibrium constant when the reaction has reached chemical equilibrium.

10.7 EFFECT OF TEMPERATURE AND PRESSURE ON EQUILIBRIUM CONSTANT

The effects of temperature and pressure on the equilibrium constant can be readily obtained by differentiating the van't Hoff isotherm with respect to the relevant parameters. First consider the effect of temperature; thus

$$\frac{d (\ln K_\alpha)}{dT} = - \frac{1}{\tilde{R}} \frac{d}{dT} \left(\frac{\Delta G^\circ}{T} \right)$$

$$= - \frac{1}{\tilde{R}} \left(\sum \frac{d}{dT} \frac{v_i \, \tilde{\mu}_i^\circ}{T} \right) = - \frac{1}{\tilde{R}} \left(-\sum v_i \frac{\tilde{h}_i^\circ}{T^2} \right)$$

where in the summation notation a positive sign is taken for the products of reaction and a negative sign for the reactants. Thus

$$\frac{d (\ln K_\alpha)}{dT} = \sum \frac{v_i \, \tilde{h}_i^\circ}{\tilde{R}T^2} = \frac{\Delta H^\circ}{\tilde{R}T^2} \tag{10.20}$$

Equation (10.20) is the well-known van't Hoff equation. It enables the equilibrium constant to be calculated at any other temperature if it is known at one temperature. The variation in ΔH° as a function of temperature is all that need be known, where ΔH° denotes the enthalpy of reaction at the standard state.

In Section 10.3 it was shown that the equilibrium constants may or may not depend on the pressure. Nevertheless, the relation governing the effect of pressure on the equilibrium constant will be derived, so that the general conditions stipulating the pressure effect can be seen. Differentiating the van't Hoff isotherm with respect to pressure yields

$$\left(\frac{\partial \ln K_{\alpha}}{\partial p}\right)_T = -\frac{1}{\check{R}T}\left(\frac{\partial \Delta G^{\circ}}{\partial p}\right)_T$$

$$= -\frac{\Delta V^{\circ}}{\check{R}T} \qquad (10.21)$$

where ΔV° represents the difference in volume of the products and reactants in their standard states.

10.8 SECOND LAW ANALYSIS OF CHEMICAL REACTIONS: WORK DONE BY CHEMICALLY REACTING SYSTEMS

The occurrence of chemical reactions is one means by which a heat reservoir may be created to produce work in a thermodynamic cycle. Heat is transferred from this reservoir and, as a result, the products are reduced to the temperature of the surroundings. The maximum work obtainable from a chemical reaction can be found from the combined form of the first and second laws for a reversible process.

Consider first a non-flow process

$$W_{\text{max}} = -(\Delta U - \int T \mathrm{d}S)$$

If the process is isothermal, this becomes

$$W_{\text{max}} = -(\Delta U - T\Delta S) = -\Delta(U - TS)_T = -\Delta A_T \qquad (10.22)$$

Similarly, for a flow process

$$W_{\text{max}} = -(\Delta H - \int T \mathrm{d}S)$$

If the process is isothermal, then

$$W_{\text{max}} = -(\Delta H - T\Delta S) = -\Delta(H - TS)_T = -\Delta G_T \qquad (10.23)$$

where the delta notation in these equations are used in the same sense as for ΔG° in equation (10.17), i.e.

$$\Delta X = \Sigma v_i \tilde{x}_i$$

where v_i refers to the coefficient of component i in the stoichiometric equation and the summation is carried out such that all output or product streams are considered positive and all input or reactant streams negative.

Thus, the maximum work obtainable from an isothermal non-flow process and an isothermal flow process is respectively the decrease in the Helmholtz and Gibbs free energies.

Since chemical reactions usually proceed during constant pressure (steady flow) and temperature processes, the Gibbs function is seen to play an even more important role than already obvious from our study of chemical equilibrium in the earlier sections.

If the constant temperature is that of the surrounding atmosphere, then equation (10.23) reduces to the form

$$W_{max,u} = - \Delta\Psi \qquad (10.24)$$

where Ψ, the availability function of the flow process is given by

$$\Psi = H - T_o S$$

Equation (10.24) can also be obtained directly by applying equation (4.11) to a steady flow reaction chamber similar to the application of the SFEE in Section 10.1.

In performing numerical calculations with equations (10.22) to (10.24), the difference in the enthalpy or internal energy terms can be taken care of using the technique introduced in conjunction with the first law in Section 10.1. To evaluate terms involving the entropy difference, a similar type of reference or standard state can be defined. Thus

$$\Delta S = \Sigma\, v_i\, \tilde{s}_i$$
$$= \Sigma\, v_i\, (\tilde{s}_i - \tilde{s}_i^o) + \Sigma\, v_i\, \tilde{s}_i^o$$

It was in the course of his investigations on this point that Nernst was led to the conclusion that the sum $\Sigma\, v_i\, \tilde{s}_i^o$ is equal to zero for a great many reactions when the standard state has a temperature of zero kelvin. This conclusion has since been enshrined as an independent law of thermodynamics—the third law. With this result, ΔS involves only a difference in entropy for each pure substance and can therefore be experimentally measured and tabulated.

10.9 THIRD LAW AND EQUILIBRIUM CONSTANTS

In the previous section, we have seen how the third law can be used to relate all measured entropy values to a single base. This result, as it turns out, also provides a convenient means for obtaining the chemical equilibrium constant, since it is often not practicable to have this quantity measured directly. The method is as follows.

From the van't Hoff isotherm

$$K_\alpha = \exp\left(- \frac{\Delta G^o}{\bar{R}T}\right)$$

Now, for an isothermal reaction

$$\Delta G^o = \Delta H^o - T\Delta S^o$$

216

The problem therefore resolves to the determination of ΔH° and ΔS°, both of which, as seen earlier, can be determined from experimental measurements.

EXAMPLES

Example 10.1   Carbon monoxide and oxygen are fed to a steady-state reaction chamber at 373.16 K, and carbon dioxide is removed at 1800 K. Find (a) the heat interaction between the reaction chamber and its surroundings per mole of carbon monoxide burned, and (b) the adiabatic temperature of the reaction. Assume the gases behave semi-perfectly and use the mean specific heats over the relevant temperature ranges for the calculations.
(a) The stoichiometric equation is

$$CO + \tfrac{1}{2}O_2 \rightleftharpoons CO_2$$

From equation (10.3)

$$\frac{Q}{\Delta \xi} = (\tilde{h}_{CD} - \tilde{h}^\circ_{CD}) - (\tilde{h}_{CM} - \tilde{h}^\circ_{CM}) - \tfrac{1}{2}(\tilde{h}_O - \tilde{h}^\circ_O) + \tilde{h}^\circ_{CD}$$

$$- \tilde{h}_{CM} - \tfrac{1}{2}\tilde{h}^\circ_O$$

where the subscripts CD, CM and O refer to CO_2, CO and O_2 respectively, for which the enthalpies of formation at 288.16 K are respectively -393.5, -110.5 and $0\ kJ/mol$.

On the assumption stated in the question, each enthalpy difference term may be written as the product of the mean specific heat and the temperature difference; thus the above equation becomes

$$\frac{Q}{\Delta \xi} = \tilde{C}_{p,CD}(1800 - 288.16) - \tilde{C}_{p,CM}(373.16 - 288.16) - \tfrac{1}{2}\tilde{C}_{p,O}(373.16 -$$

$$288.16) + (-393.5 + 110.5)$$

where it is understood that the mean specific heats over the temperature range as indicated by the temperature difference term are to be used for the respective substances. These values can be obtained from any abridged thermodynamic tables; substitution yields

$$\frac{Q}{\Delta \xi} = \frac{44 \times 1.247}{1000}(1511.8) - \frac{28 \times 1.041 - \tfrac{1}{2} \times 32 \times 0.923}{1000}(85)$$

$$- 283.0\ kJ/mol\ of\ CO\ burned$$

$$= -200.2\ kJ/mol\ of\ CO\ burned$$

(b) If the reaction is adiabatic, $Q = 0$, whence from the above solution

$$0 = \frac{44 \times 1.247}{1000}(T - 288.16) - 283.1$$

(assuming the mean $\tilde{C}_p$ between 1800 and 288.16 K for our first calculation) Solving for T yields

$$T = 5448 \text{ K}$$

Taking the next mean temperature as 2900 K, $c_{p,CD} = 1.411$ J/g whence

$$T = 4848 \text{ K}$$

On further refining the mean temperature, $c_{p,CD} = 1.399$ J/g whence

$$T = 4888 \text{ K}$$

As the next C_p value is approximately the same as the previous value, this last value is accepted as the correct adiabatic flame temperature found under the assumptions.

Remarks

(1) This example illustrates the effect of varying C_p on the adiabatic temperature of a combustion. Using C_p at room temperature will lead to too high an adiabatic temperature.
(2) More accurate values for this type of analysis can be obtained by making use of enthalpy values for semi-perfect gases tabulated in *Gas Tables* for the calculations.

Example 10.2 A stoichiometric mixture of carbon monoxide and oxygen enters a reaction chamber at 373.16 K. Carbon dioxide, carbon monoxide and oxygen leave the reaction chamber in chemical equilibrium at 2 atm and 1800 K. The equilibrium constant in terms of the partial pressures, K_p, for the reaction at 1800 K is $10^{3 \cdot 9}$ atm$^{-1/2}$. What are the partial pressures of carbon dioxide, carbon monoxide and oxygen in the products?

The stoichiometric equation for this reaction is the same as given in Example 10.1. By equation (10.7) we have

$$K_p = \frac{p_{CD}}{p_{CM}\, p_O^{1/2}} = 10^{3 \cdot 9} \text{ atm}^{-1/2} \tag{a}$$

Assuming the gases form an ideal gas mixture in the reaction chamber

$$p_{CD} + p_{CM} + p_O = 2 \tag{b}$$

and

$$n_i \propto p_i \tag{c}$$

for each gaseous substance.

By the law of conservation of atomic species for purely chemical reactions

$$n_{CD,i} + n_{CM,i} = n_{CD,f} + n_{CM,f}$$

and

$$2n_{CD,i} + n_{CM,i} + 2n_{O,i} = 2n_{CD,f} + n_{CM,f} + 2n_{O,f}$$

Now, $n_{CD,i} = 0$, since the incoming mixture consists of carbon monoxide and oxygen only, whence, for the carbon balance

$$n_{CM,i} = n_{CD,f} + n_{CM,f}$$

and

$$n_{CM,i} + 2n_{O,i} = 2n_{CD,f} + n_{CM,f} + 2n_{O,f}$$

for the oxygen balance. Substituting for $n_{CM,i}$ in the oxygen balance equation, and rearranging, gives

$$n_{O,i} = n_{O,f} + \tfrac{1}{2}n_{CD,f}$$

Since in the incoming stream, the carbon monoxide and oxygen are in stoichiometric ratio, i.e. 2 volumes of carbon monoxide to 1 volume of oxygen, it follows that

$$\frac{n_{CM,i}}{n_{O,i}} = \frac{n_{CD,f} + n_{CM,f}}{n_{O,f} + \tfrac{1}{2}n_{CD,f}} = 2 \tag{d}$$

Substitution of equation (c) into equation (d) yields

$$p_{CM} = 2p_O \tag{e}$$

where the subscript f has been dropped from equation (e) since now there is no confusion as to which pressure of the substances is referred to.

On combining equations (e) and (b)

$$p_{CD} + 3p_O = 2 \tag{f}$$

and on combining equations (e) and (a)

$$p_{CD} = 10^{3 \cdot 9}(2p_O \times p_O^{1/2}) = 16000\, p_O^{3/2} \tag{g}$$

Further combining equations (f) and (g) yields

$$p_{CD} = 16000 \left(\frac{2 - p_{CD}}{3}\right)^{1 \cdot 5}$$

$$= 7700\,(2 - p_{CD})^{1 \cdot 5} \tag{h}$$

Equation (h) can be solved by iteration to obtain the final partial pressure of carbon dioxide as follows;

assuming $p_{CD} = 1.9$ gives

$$p_{CD} = 7700\,(0.1)^{1 \cdot 5} = 243.5 \qquad \text{(too high)};$$

assuming $p_{CD} = 1.99$ gives

$$p_{CD} = 7700\,(0.01)^{1 \cdot 5} = 7.7 \qquad \text{(too high)};$$

assuming $p_{CD} = 1.999$ gives

$$p_{CD} = 7700 (0.001)^{1.5} = 0.2435 \qquad \text{(too low)}.$$

Thus the value for p_{CD} is bounded by 1.99 and 1.999 atm. Further trial shows that

$$p_{CD} = 1.996 \text{ atm}$$

From equation (f)

$$p_O = 0.00133 \text{ atm}$$

and by equation (e)

$$p_{CM} = 0.00267 \text{ atm}$$

It is therefore seen that this particular reaction is almost complete. The small amounts of carbon monoxide and oxygen remaining in the product stream are essentially due to the dissociation of the carbon dioxide at high temperatures.

Example 10.3 Determine the maximum work that can be obtained from the reaction given in Example 10.1. All heat interactions from the reaction chamber can be assumed to take place with the surroundings at a constant temperature of 300 K.

From equation (10.24)

$$
\begin{aligned}
W_{max} &= -\Delta\Psi \\
&= -\Sigma v_i \, \tilde{\psi} \\
&= \underset{R}{\Sigma} v_i \left[(\tilde{h}_i - \tilde{h}_i^o) + \tilde{h}_i^o - T_o \tilde{s}_i \right] - \underset{P}{\Sigma} v_i \left[(\tilde{h}_i - \tilde{h}_i^o) + \tilde{h}_i^o - T_o \tilde{s}_i \right]
\end{aligned}
$$

where the subscripts R and P to the summation sign refer to the reactant or incoming and product or outgoing streams respectively. Applying this expression to the reaction under consideration yields

$$
\begin{aligned}
W_{max} &= \{ (\tilde{h}_{CM} - \tilde{h}_{CM}^o) + \tilde{h}_{CM}^o - T_o \tilde{s}_{CM} \} + \tfrac{1}{2} \{ (\tilde{h}_O - \tilde{h}_O^o) + \tilde{h}_O^o - T_o \tilde{s}_O \} \\
&\quad - \{ (\tilde{h}_{CD} - \tilde{h}_{CD}^o) + \tilde{h}_{CD}^o - T_o \tilde{s}_{CD} \} \\
&= 200.2 + T_o (\tilde{s}_{CD} - \tilde{s}_{CM} - \tfrac{1}{2} \tilde{s}_O)
\end{aligned}
$$

From thermodynamic tables, the absolute entropy of the relevant substances are as follows;

$\tilde{s}_{CD}$ (at 1800 K) $= 0.3023 \text{ kJ/mol K};$
$\tilde{s}_{CM}$ (at 373.16 K) $= 0.204 \text{ kJ/mol K};$
$\tilde{s}_O$ (at 373.16 K) $= 0.2115 \text{ kJ/mol K};$

220

whence

$$W_{max} = 200.2 + 300\,(0.3023 - 0.204 - \tfrac{1}{2}0.2115)$$

$$= 200.2 - 300 \times 0.0074$$

$$= 198 \text{ kJ/mol of CO burned}$$

PROBLEMS

1. The substances P, Q and R are introduced into a steady state reaction chamber at 320 K and 2 atm. Only P and Q react chemically, R serving as a catalyst.

    The substances P_2Q and P_3Q_5 are extracted at 1100 K and 1 atm. The stoichiometric equation is

$$5P + 6Q + 10R \rightleftharpoons P_2Q + P_3Q_5 + 10R$$

Find the heat interaction with the surroundings for 1 mol of P_2Q produced. The data relevant to this reaction are as follows.

	Molecular heat capacity (kJ/mol K)	Heat of formation at 1 atm and 288 K (kJ/mol)
P	5.0	0
Q	6.0	− 5 000
R	7.0	− 10 000
P_2Q	8.0	− 50 000
P_3Q_5	9.0	− 100 000

2. Methane is burned at constant pressure with just enough oxygen to permit complete combustion. The stoichiometric equation of the reaction is

$$CH_4 + 2O_2 \rightarrow CO_2 + 2H_2O$$

If the temperature and pressure of the products of the reaction are 300 K and 1 atm respectively, find (a) the partial pressure of the water vapour in the products of combustion, (b) the mass of liquid water in the products of combustion per kg of fuel, and (c) the volume of the products of combustion per kg of fuel, neglecting the volume of the liquid water.

3. (a) Find the heat interaction from a steady state reaction chamber which receives hydrogen and theoretical air at 300 K, for complete combustion of the hydrogen supplied to steam, the products of combustion being delivered at 2500 K.

    (b) What is the adiabatic flame temperature?

4. A gaseous mixture comprising 10 mol of hydrogen, 5 mol of nitrogen and 8 mol of ammonia is in equilibrium with regard to the chemical reaction

$$N_2 + 3H_2 \rightleftharpoons 2NH_3$$

determine;

(a) the number of moles of components H_2 and N_2;
(b) the number of moles of components H_2 and NH_3;
(c) the number of moles of components N_2 and NH_3;
(d) the equilibrium constant in terms of the partial pressures for this reaction, if this equilibrium mixture can be considered as a mixture of perfect gases having a total pressure of 1 atm.

5. Show that the heat of reaction at constant pressure of a reaction between ideal gases at temperature T is related to the heat of reaction at temperature T_0 by the equation

$$\Delta h(T) = \Delta h(T_0) + \sum_{i=1}^{n} v_i \int_{T_0}^{T} c_{p_i}\, dT$$

6. Show that if in Problem 5 m of the gases are in contact with their condensed states, then

$$\Delta h(T) = \Delta h(T_0) + \sum_{i=m+1}^{n} v_i \int_{T_0}^{T} c_{p_i}\, dT + \sum_{i=1}^{m} v_i' \int_{T_0}^{T} c_{p_i}^{s}\, dT$$

where c_{p_i} denotes the heat capacity of the ith gas and $c_{p_i}^{s}$ that of the ith condensed phase.

7. One mol of argon at room temperature is heated to 20000 K at 1 atm pressure. Assume that the plasma at this condition consists of an equilibrium mixture of Ar, Ar^+, Ar^{++} and e^- according to the simultaneous reactions

$$Ar \rightleftharpoons Ar^+ + e^- \qquad (1)$$

$$Ar^+ \rightleftharpoons Ar^{++} + e^- \qquad (2)$$

The ionization equilibrium constants for these reactions at 20000 K have been determined from spectroscopic data as

$$\ln K_1 = 3.11$$
and
$$\ln K_2 = -4.92$$

Find the equilibrium composition of the plasma.

8. When a salt is dissolved in a solvent, it undergoes electrolytic dissociation until an equilibrium state has been reached. Consider the equilibrium of a binary salt S which dissociates in a solution according to the relation

$$S \rightleftharpoons S^+ + S^-$$

Deduce Ostwald's dilution law

$$\frac{\gamma^2 C}{1 - \gamma} = K, \text{ a constant at given } p, T$$

222

where γ is the degree of dissociation and C the concentration of the solute without respect to dissociation.

Why is it that this law is valid for electrolytes behaving as perfect solutions only?

CHAPTER ELEVEN

Thermodynamics of Special Systems

11.1 THERMAL RADIATION

Up to this point, thermodynamics has been shown to apply to material systems with or without mass transfers between the systems. We shall now proceed to apply the laws of thermodynamics to a system with no material mass in the conventional sense, namely, that involving only thermal radiation. The success of thermodynamics in this respect not only demonstrates its extreme generality, but also in a sense foreshadows the duality between wave and material aspects of matter. Our interest in this connection is restricted to demonstrating how the Stefan–Boltzmann law in radiation heat transfer can be duduced from pure thermodynamic reasonings. Other laws governing the behaviour of radiating bodies will be left to works dealing with radiation heat transfer.

The main concern here is a type of radiation enclosed completely by a surface which emits this radiation. This has come to be known as cavity radiation. As a prelude to the discussion, some of the relevant properties of this cavity radiation will be illustrated. Consider two cavities A and B maintained at the same temperature T and connected by a narrow tube through which radiation may pass from one to the other as shown in Figure 11.1. Now, if the radiation from A to B exceeds that from B to A, it would be possible to raise the temperature of B to $T + \Delta T$ and for B to still receive more radiation than it emits. This is contrary to the second law since experience tells us that thermal radiation is equivalent to a heat interaction. It follows, therefore, that whatever the nature of the walls of A and B the net radiation from one to the other is the same. Thus, provided the walls of the cavity are opaque, the quality of the radiation in equilibrium inside is independent of the nature of the walls and depends only on their temperature.

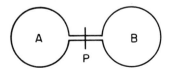

Figure 11.1

Similarly, by inserting filters or polarizers at **P**, it can be shown that the radiation is unpolarized and that the spectral distribution is independent of the walls.

Thus, the quality of cavity radiation depends on the temperature alone, and in particular, all extensive properties of the radiating system are proportional to the volume and all intensive properties are functions of temperature only.

We are now in a position to apply the combined laws of thermodynamics or the Gibbs equation to thermal radiation. The most elementary question that we have to consider first is the relevant work mode of the system. Generally, this lies outside the domain of thermodynamic reasoning and relies on some knowledge of the behaviour of the system. It turns out from studies in electromagnetic theory, or from a simple consideration akin to elementary kinetic theory of gases, viz. that the radiation is equivalent to a photon gas, that radiation does exert a pressure. Thus, the work mode for a radiating system is the same as that for a simple compressible system. The relation between the pressure and the relevant parameters is given by

$$p = \frac{1}{3} u \tag{11.1}$$

where u is the energy density. This equation may be considered in the same light as the equation of state of gaseous systems which is also independent of thermodynamic reasoning.

Substituting this into the Gibbs equation for a simple system

$$dU = TdS - pdV = TdS - \frac{1}{3} u \, dV \tag{11.2}$$

but
$$U = uV \quad \text{and} \quad S = sV$$

whence
$$dU = udV + Vdu$$

and
$$dS = sdV + Vds$$

On substituting these relations into equation (11.2)

$$du = Tds + \frac{1}{V} (Ts - \frac{4}{3} u) \, dV$$

Since u and s are functions of temperature only

$$du = Tds \tag{11.3}$$

and

$$Ts = \frac{4}{3} u \tag{11.4}$$

Hence

$$\frac{du}{u} = \frac{4}{3}\frac{ds}{s}$$

which integrates to

$$u = As^{4/3} \tag{11.5}$$

On substituting for s in terms of T from equation (11.4), equation (11.5) becomes

$$u = aT^4 \text{ and } s = \frac{4}{3}aT^3 \tag{11.6}$$

where A is a constant of integration and

$$a = \left(\frac{3}{4}\right)^4 \left(\frac{1}{A}\right)^3$$

Now the energy density in an isothermal enclosure can be found by the following considerations. For the sake of simplicity, consider the enclosure to be a sphere. The average distance the radiation emitted from any point on the sphere has to travel before it is finally absorbed by the other parts of the enclosure is given by

$$L = \frac{\int l\,dA}{\int dA}$$

where dA represents an elemental ring on the sphere at a distance l from the point of emission, as shown in Figure 11.2. But

$$l = 2r\cos\theta$$

and

$$dA = 2\pi(2r\cos\theta\sin\theta)r\,d(2\theta)$$

whence, on integration

$$L = \frac{\frac{16}{3}\pi r^3}{4\pi r^2} = \frac{4}{3}r$$

Now since the speed of the radiation is c, the average time the radiation remains in the spherical enclosure before it is absorbed is given by

$$t = \frac{4r}{3c}$$

Let E_b be the emissive power of the surface, i.e. the energy emitted per unit area of surface per unit time, then the total energy radiated into the cavity by the spherical enclosure per unit time is $4\pi r^2 E_b$, hence the energy content permeating the cavity is

$$4\pi r^2 E_b \times \frac{4r}{3c}$$

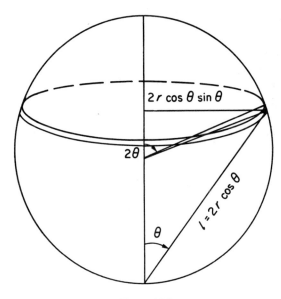

Figure 11.2

Thus the energy density is given by

$$u = \frac{\frac{16}{3c} \pi r^3 E_b}{\frac{4}{3} \pi r^3} = \frac{4 E_b}{c}$$

On substituting into equation (11.6)

$$E_b = \frac{ac}{4} T^4 = \sigma T^4 \tag{11.7}$$

which is the celebrated Stefan–Boltzmann equation. The proportional constant σ, also known as the Stefan–Boltzmann constant, can be calculated by integrating Planck's distribution over all wavelengths. Its value is 5.670×10^{-8} W/m^2 K^4.

11.2 THE FUEL CELL

A fuel cell is a device for the direct conversion of chemical energy into electric energy. It is therefore not limited by the Carnot efficiency. Basically, the structure of the fuel cell is very similar to that of a battery or an electrolytic cell in that two electrodes are placed in an electrolyte and a potential difference results from the reactions at the electrode surfaces. However, in a conventional battery the system is given an initial charge and is exhausted once this charge is gone. The fuel cell, on the other hand, remains in operation as long as fuel and oxygen is supplied to it. For the fuel cell, the electrolyte must be of such character that it permits rapid diffusion of ions through it. This diffusion process, in fact, serves as the mechanism for completing the electric circuit and thus determines the rate of electric charge flowing in the external network.

Before proceeding to a thermodynamic analysis of the fuel cell, let us first consider its work mode. The fuel cell is essentially a device operating at constant pressure, so pressure ceases to be a property describing the system. Instead, the generalized force is ε and the generalized displacement is $\mathscr{Q}$, as is evidenced from the expression for elemental work done when an elemental charge $d\mathscr{Q}$ is transported through a potential difference of ε, given by

$$dW = \varepsilon \, d\mathscr{Q}$$

Hence, the Gibbs equation for the fuel cell can be written as

$$T dS = dU + \varepsilon \, d\mathscr{Q} \tag{11.8}$$

or

$$T dS = \left(\frac{\partial U}{\partial T}\right)_{\mathscr{Q}} dT + \left[\varepsilon + \left(\frac{\partial U}{\partial \mathscr{Q}}\right)_T\right] d\mathscr{Q} \tag{11.9}$$

Using the condition that dS is a total differential gives

$$\frac{\partial}{\partial \mathscr{Q}}\left(\frac{1}{T}\frac{\partial U}{\partial T}\right)_{\mathscr{Q}}\bigg|_T = \frac{\partial}{\partial T}\left[\frac{1}{T}\left\{\varepsilon + \left(\frac{\partial U}{\partial \mathscr{Q}}\right)_T\right\}\right]_{\mathscr{Q}} \tag{11.10}$$

which yields on simplification

$$\left(\frac{\partial \varepsilon}{\partial T}\right)_{\mathscr{Q}} = \frac{1}{T}\left\{\varepsilon + \left(\frac{\partial U}{\partial \mathscr{Q}}\right)_T\right\} \tag{11.11}$$

If a specific heat $C_{\mathscr{Q}}$ is now defined by means of the equation

$$C_{\mathscr{Q}} = \left(\frac{\delta Q}{\delta T}\right)_{\mathscr{Q}} = \left(\frac{\partial U}{\partial T}\right)_{\mathscr{Q}} \tag{11.12}$$

on substituting both equations (11.11) and (11.12) into equation (11.9), the following important expression is obtained

$$T dS = C_{\mathscr{Q}} dT + T\left(\frac{\partial \varepsilon}{\partial T}\right)_{\mathscr{Q}} d\mathscr{Q} \tag{11.13}$$

If the process is isothermal, then

$$Q = T \frac{d\varepsilon}{dT} \Delta \mathcal{Q} \qquad (11.14)$$

since at near equilibrium, the emf ε is a function of temperature only.

In the meantime, the cell delivers an amount of work W_s, of magnitude given by

$$W_s = \varepsilon \Delta \mathcal{Q} \qquad (11.15)$$

If the reaction going on inside the fuel cell is assumed to go to completion, i.e. the stoichiometric equation governing the reaction is met, the electrical charge released per mole of fuel consumed will be given by Faraday's first law on electrolysis, which can be written as

$$\Delta \mathcal{Q} = nF \qquad (11.16)$$

where n is the number of chemical equivalents per mole of fuel consumed and F is a Faraday, the amount of electrical charge released by an equivalent of the reacting substance. The value of n is equal to the valence of the charge ions in the electrolyte and the value of a Faraday can be taken to be 96500 coulombs/equivalent.

Substituting equation (11.16) into equations (11.14) and (11.15)

$$Q = nFT \frac{d\varepsilon}{dT} \qquad (11.17)$$

and

$$W_s = nF\varepsilon \qquad (11.18)$$

Now considering a control volume isolating the fuel cell operating in a steady state, the SFEE gives

$$\Delta H = Q - W_s \qquad (11.19)$$

where the Δ notation denotes the difference between outlet and inlet quantities. Substituting for Q and W_s into equation (11.19) gives

$$\Delta H = nF \left(T \frac{d\varepsilon}{dT} - \varepsilon \right) = nFT^2 \frac{d}{dT} \left(\frac{\varepsilon}{T} \right) \qquad (11.20)$$

Equation (11.20) enables the determination of either the value of ε or the value of the heat of reaction from the slope of the ε/T versus T plot when the other quantity is known.

According to equation (11.20), the cell emf can be higher or lower than that corresponding to the decrease in enthalpy of the substances passing through the cell, depending on whether the variation of emf with temperature is negative or positive. From equation (11.17), this in turn corresponds to whether the reaction is endothermic or exothermic. However, as all reactions in fuel cells concern the slow oxidation of a fuel and are exothermic in nature, the cell emf is less than that given by the expression

$$\varepsilon_o = -\frac{\Delta H}{nF} \qquad (11.21)$$

The observed open circuit emf corresponds only to the maximum useful work obtainable from the reaction going on in the cell. Its value is therefore given by

$$\varepsilon_r = -\frac{\Delta G}{nF} \qquad (11.22)$$

Equation (11.22) can be derived from equation (11.19) for an isothermal change, since now $Q = T\Delta S$ and $G = \Delta(H - TS)$.

The ideal efficiency for a fuel cell is defined as the ratio of the maximum useful work $-\Delta G$, to the maximum available work $-\Delta H$; thus

$$\eta_i = \frac{\Delta G}{\Delta H} = 1 - \frac{T\Delta S}{\Delta H} = \frac{\varepsilon_r}{\varepsilon_o} \qquad (11.23)$$

If the cell is under load, the actual voltage ε_a is less than ε_r; and the actual efficiency is defined as

$$\eta_a = \frac{\varepsilon_a}{\varepsilon_o} \qquad (11.24)$$

and the effectiveness of the conversion as

$$\eta_G = \frac{\varepsilon_a}{\varepsilon_r} \qquad (11.25)$$

This last quantity is also known as the free energy efficiency. It can be readily shown that the actual efficiency of a fuel cell can be expressed as a product of the ideal and free energy efficiencies.

Furthermore, an efficiency ratio is also defined as follows

$$\eta_{ratio} = \frac{\eta_a}{\eta_i} = \frac{\varepsilon_a}{\varepsilon_r}$$

which is therefore seen to be identical to the free energy efficiency. It should be realized thus that the free energy efficiency is not an energy conversion efficiency (compare this with the isentropic efficiency of a compressor or turbine stage). It determines the amount of irreversibility due to the finite rate of the reaction in the cell. For most practical cells, η_{ratio} or η_G has values around 0.7.

Up to this point, we have analysed the fuel cell as a black box in the same way as we perform an engineering cycle analysis. Knowing that the process inside the fuel cell is one of slow combustion, further information on the cell emf can be deduced using the results of the preceding chapter.

Rewriting equation (10.19) in the form

$$- \Delta G = \tilde{R} T \ln \frac{K}{Q} \tag{11.26}$$

where ΔG represents the free energy change per mole of the substances which pass through the fuel cell, K the thermodynamic equilibrium constant and Q the reaction quotient.

Combined with equation (11.22), equation (11.26) becomes

$$\varepsilon_r = \frac{\tilde{R} T}{nF} \ln \frac{K}{Q} \tag{11.27}$$

When all reacting substances are at unit activity, $Q = 1$, and the cell is said to be in its standard state; thus

$$- \Delta G^\circ = \tilde{R} T \ln K \tag{11.28a}$$

and

$$\varepsilon_r^\circ = \frac{\tilde{R} T}{nF} \ln K \tag{11.28b}$$

where the superscript 0 refers to the standard state.

Combining the appropriate form of equation (11.28) with equation (11.27)

$$\varepsilon_r = \varepsilon_r^\circ - \frac{\tilde{R} T}{nF} \ln Q \tag{11.29}$$

where ε_r is the open circuit potential of the cell at arbitrary concentration and ε_r° the standard electrode potential for the reaction. Equation (11.29) is known as the Nernst equation. It is important that standard electrode potentials should not be confused with open circuit potentials. The former, as already shown, is the emf of the cell when each substance is in its standard state, namely, gases at one atmospheric pressure and liquids with unit activity. The latter refers to the quantity that can be readily computed via the Nernst equation, equation (11.29), and, of course, can vary with the experimental or operational conditions.

Types of fuel cell

We shall now apply the theory deduced above to several typical fuel cells. First, the various methods used in their classification will be dealt with briefly.

Fuel cells can be categorized by the manner in which fuel in consumed. According to this mode of classification, they can be divided into two main types;

(1) Direct: in which the fuel is used as an electrode;
(2) Indirect: in which the fuel is first converted into an intermediate substance such as CO before being used as an electrode.

In both types, the fuel electrode is coupled with an oxygen electrode. A cell of the first type is ideal, but unfortunately it is difficult to ionize the fuel except at very high temperatures (> 1300 K). Even so, the oxygen electrode tends to be poor in durability and performance.

Alternatively, fuel cells can be classified by their temperature of operation, as shown below.

(i) *Low temperature cells* (290–370 K)

These include carbon and carbon monoxide fuel cells, as well as cells using electrolyte soluble fuels such as methanol, hydrazine and ammonia.

For carbon fuel cells, electrodes of porous carbon and oxygen are used. The electrolyte consists of a strong alkali, such as KOH and the presence of an undisclosed catalyst is necessary to bring about electrochemical oxidation at low temperatures.

For CO fuel cells, both alkaline and acidic electrolytes can be used. For the former, carbonate solutions are generally perferred since the use of hydroxide solutions is unrealistic from the point of view of long term operation of the cell as the electrolyte will gradually be converted to carbonate anyway. A large number of catalysts have been suggested, ranging from metals in group VIA to group IB of the Periodic Table. In alkaline solutions, potentials of the CO electrode are of the order of -500 mV except for metals of group IB whose potentials are of the order of -200 mV only, all on the standard hydrogen scale. For the latter, only platinum black and Raney platinum electrodes have been used. It seems that for both type of electrolytes, CO is not oxidized directly to CO_2. Rather, hydrogen is released through certain auxiliary reactions and this is then oxidized to produce the electric current. This theory on the mechanism involved has been confirmed by the fact that the observed electrode potentials for low temperature CO cells are close to the hydrogen potential.

Cells using electrolyte soluble fuels have the advantage of simpler electrode construction. Moreover, the fuels being used are all liquids at room temperatures. This further results in easier handling and storage, and improved energy/weight and energy/volume ratios because of the higher calorific value per unit volume of liquid fuels as compared to their gaseous counterparts. Of the three fuels suggested, hydrazine is the most easily oxidized and hence has the greatest power output. It is, however, mildly caustic, quite toxic and very expensive, and it is unlikely that its cost will come down sufficiently even if the demand were to increase substantially. Moreover, hydrazine has the lowest theoretical energy content if it has to be stored as the hydrate, thus entailing a cell of greater weight than the others for the same energy output. The inherent simplicity of these cells makes them very attractive from an engineering stand-point. It is therefore expected that cells using the cheaper fuels can enjoy a wide market if they can be evolved in the near future.

(ii) Medium temperature cells (400–500 K)

These include hydrogen and hydrocarbon fuel cells using aqueous electro-lytes. For hydrogen fuel cells, KOH is in many ways the best electrolyte, though the performance of hydrogen electrodes appears to be better in acid than in alkaline electrolytes. This is because the wide choice of relatively cheap metals that may then be used for constructing the cell eases the engineering problems. In actual fact, the operating temperature of hydrogen cells depends on the type of catalyst used. Noble metals required temperatures only slightly higher than room temperatures for their operation, whilst higher temperatures are needed to accelerate the reaction in cells using nickel or Raney nickel-aluminium alloy, though more recently, a semiconductor, nickel boride, has been reported to be almost as active as palladium for the electrochemical oxidation of hydrogen.

For hydrocarbon fuel cells, platinum black has been found to be the best simple electro-catalyst for the electrochemical oxidation of hydrocarbons at temperatures of less than 500 K. The nature of the electrolyte, however, strongly affects the rate of oxidation: most hydrocarbons show greatest activity in strongly acid media. Further advantages of acid electrolytes include the re-jection of CO_2 and the higher conductivities obtained than with other aqueous media. The acid found most suitable for use as electrolyte is orthophosphoric acid since sulphuric acid has the tendency to be directly reduced by the hydro-carbon fuel at temperatures of around 425–475 K. However, material problems facing this type of cells are quite tremendous because (i) phosporic acid is ex-tremely corrosive at these temperatures and (ii) the catalyst used is very ex-pensive.

(iii) High temperature cells (570–1070 K)

These include cells using molten carbonates and solid oxide electrolytes. With fused carbonates, the CO_2 obtained from the reaction can be withdrawn from the anodic products and fed into the air or oxygen stream at the cathode to maintain the electrolyte invariant, while other salts used would gradually be converted into the carbonate also. The use of oxides simplifies the mechanisms involved since the oxygen can be directly ionized at the cathode to replenish the oxygen ions.

For carbonate electrolytes, with hydrogen as fuel, the reactions at the electro-des are:

cathode: $\quad O_2 + 2CO_2 + 4e^- \rightleftharpoons 2CO_3^=$ $\qquad$ (11.30)

anode: $\quad 2H_2 + 2CO_3^= \quad\rightleftharpoons 2H_2O + 2CO_2 + 4e^-$ $\qquad$ (11.31)

overall: $\quad O_2 + 2CO_2 + 2H_2 \rightarrow 2H_2O + 2CO_2$ $\qquad$ (11.32)

Thus, the transfer of oxygen from air to the fuel electrode proceeds as a transfer of carbonate ions.

The electrolytes used in practical cells are usually mixtures of either sodium and lithium or sodium, lithium and potassium carbonates. These mixtures enable the cell to operate in the temperature range of 660–1070 K. At these temperatures, it is unlikely that hydrocarbons are utilized directly at the fuel electrode. Oxidation probably occurs by an indirect mechanism involving the following processes, illustrated for the case of ethane as fuel:

thermal cracking: $\qquad C_2H_6 \qquad \rightleftharpoons C_2H_4 + H_2$ (11.33)

steam reforming: $\qquad C_2H_4 + 2H_2O \rightleftharpoons 2CO + 4H_2$ (11.34)

water gas shift: $\qquad CO + H_2O \rightleftharpoons CO_2 + H_2$ (11.35)

followed by the electrochemical oxidation of the hydrogen

$$H_2 + O^= \rightleftharpoons H_2O + 2e^- \qquad (11.36)$$

The overall results of this sequence is the same as would be obtained by direct oxidation of ethane at the fuel electrode

$$C_2H_6 + 7O^= \rightarrow 2CO_2 + 3H_2O + 14e^- \quad (11.37)$$

Such a theory has been substantiated by the fact that the open circuit voltages are close to the theoretical values with hydrogen as fuel. Even without this series of reactions, hydrogen still constitutes in essence the fuel consumed at the cathode considering the fact that many types of practical cells using hydrocarbon fuels use fuels reformed with steam and/or carbon dioxide, to avoid carbon deposition in the cell as a result of the thermal cracking of the fuel in the cell.

For solid oxide cells, zirconia, stabilized by other oxides, such as calcia or yttria, having a cubic structure and which forms solutions with zirconia, has been found to be the most practical. In many respects, solid oxide cells are similar to carbonate ones. If a reduction in the operating temperatures of these cells from the present 1273 K to 973 K or so could be achieved, the greater simplicity of the oxide cells would probably ensure their ultimate sucess at the expense of the fused carbonate cells.

Since high temperature cells give rise to a host of engineering problems, one may well be tempted to ask why we want such a high operating temperature if fuel cells can also be built to run at low or medium temperatures. The main advantage of a high temperature cell is that it can be operated without a catalyst, which is a distinct advantage when it is realized that most catalysts are noble metals of exceedingly high cost and limited availability. The general absence of activation polarization in these cells seems to bear out this claim. The choice of electrode materials is, however, not as wide as was once envisaged, since corrosion of the electrode materials in the molten carbonate electrolyte has often been found to be a limiting factor.

234

Losses in fuel cells

We have seen earlier that the open circuit or zero current potential of a fuel cell is higher than the ideal cell voltage at finite output. This drop in cell voltage at increased current is caused by a number of losses, which can be neatly summarized in the form of a voltage/current–density diagram as shown in Figure 11.3.

When power is drawn from the cell, the terminal voltage decreases from

$$\varepsilon_o = -\frac{\Delta H}{nF}$$

to lower than

$$\varepsilon_r = -\frac{\Delta G}{nF}$$

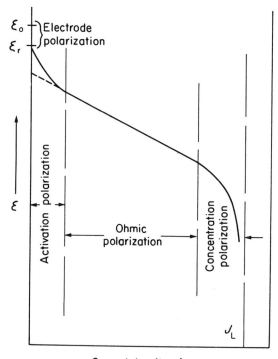

Current density J

Figure 11.3

This decrease in cell voltage reflects not only the $I^2 R$ drop within the cell, but also potential losses associated with the irreversibility of various processes at the electrode–electrolyte interface. This drop in potential has been called *electrode polarization*.

The reversibility of a given electrode reaction, and hence the electrode polarization are dependent on the catalytic properties of the electrode surface. Catalysts are used to increase the rate of the reactions at the electrodes, and the result is an increase in entropy due to irreversibility and a corresponding decrease in voltage.

Next comes what is known as *activation polarization*, the range of which is due to the rate processes that governs the transformation of fuel molecules from the neutral to an oxidized state. The difficulty of ionization at low current densities results in this potential loss. This is a surface phenomenon. Electrochemical oxidation of fuel involves a chemisorption process. The fuel molecules adsorbed on a surface in the presence of a catalyst become deprived of electrons and thus become charged. This procedure requires a certain amount of activation energy which has to be deducted from the output.

The *ohmic polarization* is attributable to the resistances of the electrolyte and electrode to the transfer of electrons or ions. This loss appears as an undesirable heating of the system.

At high current densities, *concentration polarization* arises where ion concentrations have been lowered because of difficulties in mass transport, such as passing ions through an electrolyte, supplying reactants and removing products from the electolyte–gas interface. This imposes a limiting current density J_L on the cell. This limit is affected by the reactivity requirements imposed by the given design. It is desirable to have a large gas–electrolyte interface as well as high temperature and pressure of operation.

Problems associated with actual fuel cells

Before leaving this topic, we shall summarize briefly the major problems faced in the design and construction of actual fuel cells. These are as follows.

(i) The fuel must be transported to the electrode surface in contact with the electrolyte. Since this transport process involves the diffusion at the surface of the electrode, proper selection and design of porous electrode/ materials are essential in order that losses can be minimized.

(ii) The products of reaction must be transported away from their respective electrodes. Again, complicated diffusion processes are involved and losses are encountered.

(iii) The electrolyte must be selected so that the proper ion carriers are provided for. Acid electrolytes, for example, could provide H^+ ions and alkaline electrolytes OH^- ions. An additional restriction on the electrolyte is that it should not be corrosive to the electrodes or dissolve both oxygen and fuel, so that unwanted reactions should not occur when the cell is not in use.

(iv) Electrodes must be selected which do not impede the electron transport at the surface during fuel–electrolyte–electrode interaction process.

11.3   INCLUSION OF SURFACE EFFECTS

So far, in our discussion on systems consisting of two or more phases separated by a surface, the effect of the surface has been neglected. However, since the surface effects play an important part in certain thermodynamic systems, we shall proceed to show how this can be incorporated into our analysis. From elementary physics, we know that the work done against surface tension is given by $\sigma \, d\mathscr{A}$ where σ is the surface tension and $d\mathscr{A}$ the change in area of the system. Thus, for a compressible system for which the effect of surface tension has to be included, two work modes are possible and the Gibbs equation becomes

$$T dS = dU + p dV - \sigma \, d\mathscr{A} \quad \text{or} \quad dU = T dS - p dV + \sigma \, d\mathscr{A} \quad (11.38)$$

In this relation, the internal energy of the system can be considered as a function of the variables S, V and $\mathscr{A}$. In terms of T, p and $\mathscr{A}$ as independent variables, we can rewrite equation (11.38) as

$$
\begin{aligned}
dU &= \left[T \left(\frac{\partial S}{\partial T} \right)_{p,\mathscr{A}} - p \left(\frac{\partial V}{\partial T} \right)_{p,\mathscr{A}} \right] dT + \left[T \left(\frac{\partial S}{\partial p} \right)_{T,\mathscr{A}} - p \left(\frac{\partial V}{\partial p} \right)_{T,\mathscr{A}} \right] dp \\
&+ \left[\sigma + T \left(\frac{\partial S}{\partial \mathscr{A}} \right)_{p,T} - p \left(\frac{\partial V}{\partial \mathscr{A}} \right)_{p,T} \right] d\mathscr{A}
\end{aligned}
\quad (11.39)
$$

Let us proceed to give a physical interpretation to the various terms in equation (11.39). The two dT terms are easily interpretable: the first is merely the thermal capacity at constant p and $\mathscr{A}$, while the second represents the mechanical work done by the hydrostatic pressure during thermal expansion. In a similar vein, the second dp term represents the mechanical work done by the hydrostatic pressure due to the bulk elasticity of the system under pressure. The remaining terms are less easily interpreted directly and it is helpful to first transform them by means of the Marxwell relations. Subjecting the first two terms of the second form of equation (11.38) to a Legendre transformation

$$
\begin{aligned}
dG' &= dU - d(TS) + d(pV) \\
&= - S dT + V dp + \sigma \, d\mathscr{A}
\end{aligned}
\quad (11.40)
$$

Applying the condition that dG' is exact, the following three Maxwell relations are obtained

$$
\left(\frac{\partial S}{\partial p} \right)_{T,\mathscr{A}} = - \left(\frac{\partial V}{\partial T} \right)_{p,\mathscr{A}}
$$

$$
\left(\frac{\partial S}{\partial \mathscr{A}} \right)_{p,T} = - \left(\frac{\partial \sigma}{\partial T} \right)_{p,\mathscr{A}} \quad (11.41)
$$

$$
\left(\frac{\partial V}{\partial \mathscr{A}} \right)_{p,T} = \left(\frac{\partial \sigma}{\partial p} \right)_{T,\mathscr{A}}
$$

The remaining twenty-one analogues, which are of no concern here, may be derived by similar methods. With the above relations, interpretation of the remaining differential coefficients in equation (11.39) can be resumed. The first dp term is now seen to be related also to the coefficient of thermal expansion of the system, while the last two terms of the $d\mathscr{A}$ group are due to the temperature and pressure coefficients of the surface tension respectively.

When the temperature and pressure of the system remain constant, equation (11.39) becomes

$$dU_{p,T} = \left[\sigma - T\left(\frac{\partial \sigma}{\partial T}\right)_{p,\mathscr{A}} - p\left(\frac{\partial V}{\partial \mathscr{A}}\right)_{p,T} \right] d\mathscr{A} \qquad (11.42)$$

The significance of the last term of the above expression is of considerable interest. At first thought, one would not expect the volume of the liquid to depend on the surface area of the system, i.e., the coefficient $(\partial V/\partial \mathscr{A})_{p,T}$ should equal zero. However, this coefficient might be different from zero if the average volume occupied by the molecules near the surface of the film is different from that in the interior. If $(\partial V/\partial \mathscr{A})_{p,T} < 0$, then the density of molecules must be greater nearer the surface. This is because as the surface film is expanded its volume decreases and hence its average density increases. Increasing density near the surface is known as surface adsorption. In this case, it follows from the Maxwell relations that the surface tension decreases as pressure increases. When adsorption effects are neglected, equation (11.42) becomes

$$dU_{p,T} = \left[\sigma - T\left(\frac{\partial \sigma}{\partial T}\right)_{p,\mathscr{A}} \right] d\mathscr{A} \qquad (11.43)$$

Dividing through by $d\mathscr{A}$, and defining the differential coefficient $(\partial U/\partial \mathscr{A})_{p,T}$ as the surface energy of the system—denoted by U_s, gives

$$U_s = \sigma - T\left(\frac{\partial \sigma}{\partial T}\right)_{p,\mathscr{A}} = \sigma - T\frac{d\sigma}{dT} \qquad (11.44)$$

since now σ does not depend on p and not at all on $\mathscr{A}$.

Following the same approach, a surface entropy can be defined for the system, and the second relation of equation (11.41) can now be written as

$$S_s = -\frac{d\sigma}{dT} \qquad (11.45)$$

This when combined with equation (11.44) gives

$$U_s = \sigma + TS_s$$

or

$$\sigma = U_s - TS_s \tag{11.46}$$

which indicates that σ can be interpreted as the surface Helmholtz free energy of the system.

In order to actually evaluate the various surface properties of the system, an 'equation of state' for the variables σ, T, p and $\mathscr{A}$ must be introduced at this stage. As mentioned earlier, σ is independent of p and $\mathscr{A}$ for nonadsorbing systems. It has been found that the surface tension of various pure liquids shows a trend which can be rather closely represented by the empirical relation of Ferguson

$$\sigma = \sigma_0 (1 - T_r)^{11/9} \tag{11.47}$$

where T_r is the reduced temperature and σ_0 an empirical constant which may be regarded as the surface tension of a hypothetical subcooled liquid at 0 K. When equation (11.47) holds, the surface entropy is

$$S_s = \frac{11}{9} \frac{\sigma_0}{T_c} (1 - T_r)^{2/9} \tag{11.48}$$

U_s can now be computed using both equations (11.47) and (11.48).

Consider now the conditions for equilibrium between two phases of a single component system separated by a surface whose effects are not negligible. We may visualize this to be a spherical drop immersed in its own vapour. Elementary physics tells us that for mechanical equilibrium the pressure across the surface of the drop are not equal, i.e. p is not uniform throughout the system. T and V must therefore be chosen as independent variables and the Helmholtz free energy used to establish the conditions of equilibrium for the system.

The total Helmholtz free energy of the system is given by

$$A = A_f + A_g + A_s = a_f m_f + a_g m_g + A_s \tag{11.49}$$

where the subscripts f, g and s refer to the liquid, gaseous and surface phases respectively. Also, we must have for a closed system

and

$$dm_f + dm_g = 0 \tag{11.50}$$

or

$$d(m_f v_f) + d(m_g v_g) = 0$$

$$(v_f - v_g) dm_f = - m_f dv_f - m_g dv_g \tag{11.51}$$

Now for a spherical drop

$$\frac{dV}{d\mathscr{A}} = \frac{r}{2}$$

where r is the radius of the drop. But $V = m_f v_f$, whence

$$d\mathscr{A} = \frac{2}{r}(m_f dv_f + v_f dm_f) \tag{11.52}$$

Since the surface film does not constitute an adiabatic wall, it follows therefore that at equilibrium the temperature of the coexisting phases must be equal. Now, for any changes between equilibrium states of the system, it is necessary that these changes be reversible isothermal processes. For such processes, the reversible work interaction (represented by the change in the Helmholtz free energy) with the surroundings is zero. Thus

$$dA = a_f dm_f + a_g dm_g + m_f da_f + m_g da_g + dA_s = 0 \tag{11.53}$$

Now for the liquid and vapour phases

$$da = -s\,dT - p\,dv \tag{11.54}$$

and the analogous relation for the surface phase is

$$dA_s = -S_s dT + \sigma\,d\mathscr{A} \tag{11.55}$$

Substituting equations (11.54) and (11.55) into equation (11.53) and noting that $dT = 0$, gives

$$a_f dm_f + a_g dm_g - m_f p_f dv_f - m_g p_g dv_g + \sigma\,d\mathscr{A} = 0$$

On combining with equations (11.50) and (11.52), this becomes

$$\left[a_f - a_g + \frac{2\sigma v}{r} \right] dm_f - m_f p_f dv_f - m_g p_g dv_g + \frac{2\sigma m_f}{r} dv_f = 0 \tag{11.56}$$

On combining with equation (11.51), (11.56) becomes

$$\left[a_f - a_g + \frac{2\sigma v_f}{r} \right] \frac{m_f dv_f + m_g dv_g}{v_f - v_g} + \left[m_f p_f - \frac{2\sigma m_f}{r} \right] dv_f$$

$$+ m_g p_g\,dv_g = 0 \tag{11.57}$$

Now dv_f and dv_g are independently variable parameters, so on equating their respective coefficients to zero the following expression is obtained

$$\frac{a_f - a_g + \frac{2\sigma v_1}{r}}{v_f - v_g} = -p_f + \frac{2\sigma}{r} = -p_g \tag{11.58}$$

whence

$$\Delta p = p_f - p_g = \frac{2\sigma}{r} \tag{11.59}$$

a well known result in elementary physics for the mechanical equilibrium of the drop. Note that this has been arrived at in the above discussion by purely thermodynamic arguments! Also

$$a_f - a_g + p_g(v_f - v_g) = -\frac{2\sigma v_f}{r} = -(p_f - p_g)v_f$$

i.e.

$$(a_f + p_f v_f) - (a_g + p_g v_g) = 0$$

or

$$g_f(T, p + \Delta p) = g_g(T, p) \tag{11.60}$$

Thus, the condition for equilibrium is still that the Gibbs free energy of each phase remains equal; however, they now refer to different pressures.

Equation (11.60) may be used to find the effect of surface tension on vapour pressure. Differentiating this with respect to r at constant temperature

$$\left(\frac{\partial g_f}{\partial p_f}\right)_T \left(\frac{dp_f}{dr}\right) = \left(\frac{\partial g_g}{\partial p_g}\right)_T \frac{dp_g}{dr}$$

i.e.

$$v_f\left(dp_g - \frac{2\sigma}{r^2}\,dr\right) = v_g\,dp_g$$

$$(v_g - v_f)\,dp_g = -\frac{2\sigma}{r^2}v_f\,dr \tag{11.61}$$

If v_f is negligible compared to v_g and assuming that the liquid is incompressible and the vapour behaves as a perfect gas, equation (11.61) becomes

$$\frac{\tilde{R}T}{Mp_g}\,dp_g = -\frac{2\sigma}{r^2}v_f\,dr$$

where $\tilde{R}$ is the universal gas constant and M the molecular weight of the vapour. On integrating, this yields

$$\frac{\tilde{R}T}{M}\ln\left(\frac{p}{p_o}\right) = \frac{2\sigma}{r}v_f$$

or

$$p = p_o \exp\left(2\sigma\, v_f\, M/r\tilde{R}T\right) \tag{11.62}$$

where p_0 is the vapour pressure over a plane liquid surface (i.e. for $r \to \infty$). Equation (11.62) is known as the Kelvin–Helmholtz relation. If the change in vapour pressure is small, the exponential term may be expanded to give

$$\frac{\Delta p}{p_0} = \frac{2\sigma \, v_f M}{r \tilde{R} T} \tag{11.63}$$

Generally, the effect of surface tension on vapour pressure only becomes important when the radius of curvature of the surface is very small. For water, for example, at room temperature of about 300 K, $\sigma = 0.072$ N/m and $v_f = 0.001$ m^3/kg; using the specific gas constant for steam of 0.4615 kJ/kg, equation (11.63) gives

$$\Delta p/p_0 \simeq 10^{-3}/r$$

where r is in microns (μm). Thus, a drop of 0.1 μm in radius has its vapour pressure increased by only 1 percent. However, the modification of the vapour pressure does become a crucial factor in processes involving the nucleation of one phase within another. Nevertheless, further discussions on this topic will be left to more specialized texts.

Before leaving this topic, we shall briefly touch on the instability of the bubble once it is formed. Some indication of this can be obtained by a more detailed examination of the first relation of equation (11.62). It is seen that at the same external pressure the small bubble always has a greater pressure and hence chemical potential than the bulk liquid. Thus, vapour will escape from the bubble to the liquid and as the bubble becomes smaller its pressure and hence chemical potential increases still more until eventually the bubble will completely disappear. To get a more realistic description of the situation, a more elaborate model must be considered. As an illustration, consider a bubble immersed in a thin layer of liquid exposed to the atmosphere of pressure p_0. The condition for equilibrium of the bubble is

$$p_v = p_0 + \frac{2\sigma}{r} \tag{11.64}$$

On the basis of this relation, it appears that without suitable nuclei in the liquid, small bubbles could never be self initiated since it would require a prohibitively large vapour pressure to induce the process of bubble growth. To simplify the discussion, assume that nuclei exist in the liquid through dissolved gases which are driven out of solution as the liquid is heated. Suppose now that a gas bubble of radius a exists in the liquid at pressure p_0. If vapour evaporates into the bubble causing it to grow to a new radius r, then the total pressure in the bubble becomes

$$p_v + p_0 \frac{a^3}{r^3}$$

As this is the total pressure internal to the bubble, the condition for equilibrium now becomes

$$p_v + p_o \frac{a^3}{r^3} = p_o + \frac{2\sigma}{r}$$

or

$$p_v - p_o = \frac{2\sigma}{r} - p_o \frac{a^3}{r^3} \qquad (11.65)$$

The form of equation (11.65) indicates that the excess pressure has a maximum value. It can be easily shown that this occurs at a value of r given by

$$r_m^2 = 3 p_o a^3 / 2\sigma \qquad (11.66)$$

Substituting this into equation (11.65) the maximum excess pressure is given by

$$p_v - p_o = \frac{4\sigma}{3 r_m} \qquad (11.67)$$

On the basis of the above analysis, the vapour pressure required to initiate nucleate boiling is a function of the radius of the nucleus, the atmospheric pressure, and the surface tension of the liquid with its vapour. Once p_v reaches the value given by equation (11.67), the bubble continues to grow, in other words, the liquid boils. It is this pressure excess which leads to boiling at temperatures in excess of the saturation temperature for liquids containing only minute gas bubbles. For water, to raise the boiling point by 1 K under atmospheric pressure requires about a 3 percent increase in the vapour pressure. For this amount of excess pressure bubbles of radius not greater than 0.03 μm would be required.

11.4 THERMODYNAMICS OF GENERAL SYSTEMS

It was seen in the earlier sections of this chapter that the procedure adopted for analysing systems other than the simple compressible system is essentially similar. Firstly, in Section 11.1 it was shown how the problem of a simple system (meaning one possessing a single work mode only), qualitatively different from the compressible system, can be tackled. Then, in Section 11.2 it was shown how a simple system involving a new type of work can be described. The essential steps in both cases are seen to be: first to give a new work term to the system, second to incorporate this new work term into the Gibbs equation, and third to deduce some useful results using a new equation of state describing the behaviour of the system. The formality of the exposition for both cases are identical, except in the details governing how the new work term is to be computed as well as the theoretical reasoning behind the equation of state which was left out of the discussion. In Section 11.3, analysis of a compound system was introduced, namely, one involving more than one work mode; the complications that may result were barely touched on. However, it is again obvious that the formality of the thermodynamic description remains unchanged, only that the mathematics become more involved. In the present section, these results will be generalized into a form suitable for describing any arbitrary system no matter how complex and a summary of the fundamental results will be given.

The results encountered so far for the term δW are typical of the form it takes in a reversible change. On this intuitive basis, we may represent the reversible work term by the product of a generalized force X and a generalized infinitesimal displacement dx so that the Gibbs equation can be written for reversible changes as

$$dU = T\,dS - \sum_i X_i\,dx_i \qquad (11.68)$$

As this equation portrays a relation among only properties of the system, a thermodynamic description of the system can be achieved by means of formal mathematics—step 2 of the above scheme. However, to enable us to carry out this step, the various generalized forces and displacements that are pertinent to the description must first be identified. Finally, to allow numerical computation to be carried out with these properties, an expression is needed to relate several easily measured parameters of the system—this being our equation of state. In the following table, a summary is given of the various simple systems that have come under thermodynamic study; their work terms are identified and, wherever possible, the more commonly used equations of state for these simple systems are stated.

EXAMPLE

Example 11.1 It is found in the operation of a hydrogen–oxygen fuel cell that a heat interaction of -48 kJ/mol is required between the cell and its surroundings in order to maintain the cell in an isothermal state at 298 K. Find the cell emf at this temperature.

The net effect of the reactions in the cell is

$$H_2 + \tfrac{1}{2}O_2 \rightarrow H_2O$$

for which

$$\Delta\tilde{h} = -286 \text{ kJ/mol}$$

at 1 atm and 298 K. (The student should check this result using the method of chapter 10.)

From equations (11.18) and (11.19)

$$nF\varepsilon = \mathcal{Q} - \Delta\tilde{h}$$

$$= -48 + 286 = 238 \text{ kJ/mol}$$

Table 11.1 Characteristics of various simple systems

System	Generalized force		Generalized displacement		Work done	Equation of state
Compressible	Pressure	p	Change in volume	dv	$p\,dv$	$pv = RT$ (ideal gas law)
Elastic	Stress	σ	Strain	$d\epsilon$	$-\sigma\,d\epsilon$	$\sigma = E\epsilon$ (Hooke's law; $E =$ Young's modulus)
Surface film	Surface tension	σ	Area	$d\mathcal{A}$	$-\sigma\,d\mathcal{A}$	$\sigma = \sigma_0(1 - T_r)^{11/9}$ (Ferguson's law)
Stretched wire	Tension	F	Extension	$d\ell$	$-F\,d\ell$	$F = k\ell$ (Hooke's law)
Reversible cell	Cell emf	ε	Charge	$d\mathcal{Q}$	$-\varepsilon\,d\mathcal{Q}$	$d\mathcal{Q} = F\,dn$ (Faraday's law)
Capacitor	Voltage	$\mathcal{V}$	Charge	$d\mathcal{Q}$	$-\mathcal{V}\,d\mathcal{Q}$	$d\mathcal{Q} = F\,dn$ (Faraday's law)
Dielectric	Applied field	E	Polarization	$d\mathcal{P}$	$-E\,d\mathcal{P}$	$\dfrac{\mathcal{P}}{V} = \left(a + \dfrac{b}{T}\right)E$ (a, b are constants; $V =$ volume of dielectric)
Magnetic	Magnetic field	$\mathcal{H}$	Magnetization	$d\mathcal{M}$	$-\mathcal{H}\,d\mathcal{M}$	Paramagnetic Materials: $\mathcal{M} = \dfrac{a}{T}\,\mathcal{H}$ (Curie's law; a is constant) Ferromagnetic materials: $\mathcal{M} = \dfrac{C}{T - \Theta}\,\mathcal{H}$ (Curie–Weiss law; C, Θ are constants)

whence

$$\varepsilon = \frac{238000}{2 \times 96500}\ \frac{\text{J/mol}}{\text{C/mol}}$$

$$= 1.23\ \text{V}$$

PROBLEMS

1. The radiation energy flux from the sun is about 1410 W/m² near the earth. Estimate the radiation pressure and the area of a 'solar sail' required to develop 1 N of thrust.

2. Not all of the sun's energy radiated on to the earth's surface can be utilized by a solar power plant because of the re-radiation of energy back into space when the collector surface becomes heated up. Assuming the collector to be a black body, find the maximum surface temperature, i.e. the temperature at which all incident radiation is re-radiated back into space. Assuming the

power plant operates at Carnot efficiency and rejects heat to the surroundings at a temperature of T_o, find the collector surface temperature T_c, for which the plant has maximum power output.

What is the efficiency of the plant if $T_o = 300$ K. Also find the collector area that would produce 1 kW of shaft power.

3. What effect does the use of a non-black collector surface have on the maximum power plant? (i.e. resolve Problem 2 assuming the re-radiation phenomenon to be governed by $\mathscr{Q} = \epsilon \sigma A T^4$, where ϵ is less than one; and investigate the effect of ϵ on the maximum power output.)

4. In a hydrogen–oxygen fuel cell operating at a steady state, the following reactions occur.

anode: $\qquad\qquad\qquad\qquad H_2 \rightarrow 2H^+ + 2e^-$

cathode: $\qquad \tfrac{1}{2}O_2 + 2H^+ + 2e^- \rightarrow H_2O$

overall: $\qquad\qquad\qquad\qquad H_2 + \tfrac{1}{2}O_2 \rightarrow H_2O$

In a test at 473.16 K, the following observations were recorded:

current density (A/m^2)	100	1000	2510	5030	6810
voltage (V)	1.02	0.905	0.805	0.677	0.585

For each of the tests given find the following;

(a) power generated per m^2;
(b) rate of consumption of fuel per m^2;
(c) heat interaction per m^2 of the electrodes;
(d) the maximum voltage of a hydrogen–oxygen fuel cell (H^+ carries an electric charge of 96500 coulomb/mol);
(e) the irreversibility of the cell for each of the test conditions.

5. Find the radius of the largest drop of water that can evaporate at 273.16 K without cooling and without having heat supplied to it. (The surface tension of water at 273.16 K is 0.67600 N/m and decreases by 0.15×10^{-3} N/m for each kelvin the temperature is raised. The latent heat of evaporation of water at 273.16 K is 2500 kJ/kg.)

6. By expressing the heat interaction during the reversible charging of a condenser in the form

$$\delta Q = dU - \mathscr{V}d\mathscr{Q}$$

where U is a function of temperature and charge $\mathscr{Q}$, show that the entropy change may be expressed as

$$T dS = \left(\frac{\partial U}{\partial T}\right)_{\mathscr{Q}} dT + \frac{\mathscr{Q}T}{C^2}\frac{dC}{dT}d\mathscr{Q}$$

246

where C denotes the capacity of the condenser, which is a function of T only. Hence show that the increase of entropy on charging isothermally is given by

$$\frac{1}{2}\left(\frac{2}{C}\right)^2 \frac{dC}{dT}$$

7. Derive the analogue of the Maxwell relations for a simple paramagnetic substance.

8. Show that the heat ΔQ supplied and the entropy change ΔS during the isothermal magnetization of a body are given by

$$\Delta Q = T\Delta S = T \int \left(\frac{\partial \mathcal{M}}{\partial T}\right)_{\mathcal{H}} \cdot d\mathcal{H}$$

Further show that if $\mathcal{M}$ is a function of $\mathcal{H}/T$, the quantities ΔQ, and ΔS become

$$\Delta Q = T\Delta S = - \int \mathcal{H} \, d\mathcal{M}$$

Hence, find the entropy change during an isothermal magnetization process for a Curie substance.

9. Show that

$$\left(\frac{\partial S}{\partial \mathcal{H}}\right)_T = \left(\frac{\partial \mathcal{M}}{\partial T}\right)_{\mathcal{H}}$$

$$\left(\frac{\partial T}{\partial \mathcal{H}}\right)_S = - \left(\frac{\partial \mathcal{M}}{\partial S}\right)_{\mathcal{H}}$$

CHAPTER TWELVE

Introduction to Thermodynamics of Irreversible Processes

12.1 INTRODUCTION

The main concern of classical thermodynamics, as seen earlier, is the study of systems in equilibrium and quasistatic states and to relate the thermodynamic properties associated with these states. It deals adequately with processes which begin and end with equilibrium states, even though the intermediate states of the process may be nonequilibrium and nonquasistatic. However, classical thermodynamics provides no information on the rate at which a process takes place. Since real or irreversible processes occur at finite rates, the rate at which the process occurs is usually a matter of significant interest. Its prediction for individual processes is usually treated apart from thermodynamics under the title *Transport Processes*. In this chapter, an introduction to the underlying principles of these processes will be provided and a broad basis outlined for the treatment of any situation where the processes are coupled together.

The present discussion will be necessarily restricted to the treatment of only one-dimensional systems. However, this treatment can readily be extended to spatial systems using vector notation.

12.2 PHENOMENOLOGICAL LAWS AND ONSAGER'S RECIPROCAL RELATIONS

Irreversibie processes are usually associated with the transport of one or more of the following quantities: heat, mass, momentum and electric charge. In each case, there is a *flux* of the transported quantity and producing this flux is a *driving force*, usually described by the gradient of some physical property of the system. For instance, a temperature gradient is the driving force for a heat flux, a concentration gradient for a mass flux and a gradient in electric potential for an electric current. The flux is defined as the flow per unit area of the substance involved. For uncoupled flows, it has been observed that the fluxes are proportional to their respective driving forces. Hence, the general transport phenomenon for a one-dimensional system can be written as

$$J = LX \tag{12.1}$$

where J is the flux of the transported quantity along the direction considered, L is the proportionality function, generally known as the transport coefficient and X is the driving force or gradient causing the flux in the direction under consideration

The relations for each of the individual transport processes are as follows.
Heat transfer (Fourier's law)

$$J_Q = -k \frac{dT}{dy} = -a \frac{d(\rho c_p T)}{dy}$$

Momentum transfer (Newton's law)

$$J_M = -\mu \frac{du}{dy} = -v \frac{d(\rho u)}{dy}$$

Mass transfer (Fick's law)

$$J_m = -D \frac{dc}{dy}$$

Flow of electricity (Ohm's law)

$$J_e = -\lambda \frac{dE}{dy}$$

It should be fairly obvious that these relations are really not 'laws' in the same sense of the laws of thermodynamics, but merely phenomenological equations describing the physical transport process which also serve as defining relations for the proportionality functions k, μ, D and λ, which are usually strong functions of material properties.

If only one driving force is present in a system, we say that there is a *single flow* associated with the property gradient, which may be described by the same type of phenomenological relation as those listed above. For a single irreversible flow, the technique developed in classical thermodynamics plus its appropriate phenomenological relation suffice for its analysis. The same may also be said of two simultaneous irreversible flows, provided that each is entirely independent of the other. However, in actual situations, simultaneous irreversible processes in the same system will not be independent of each other's potential gradient. Under such conditions, we say that we have *coupled flows*. It is for tackling such situations that the techniques of the thermodynamics of irreversible processes are developed.

Provided that the gradients X_i are not too great, the fluxes J_i will generally be linear functions of the driving forces. This may be expressed as

$$J_i = L_{i1} X_i + L_{i2} X_2 + \ldots + L_{in} X_n \tag{12.2}$$

where $i = 1, \ldots, n$.

These relations are called linear phenomenological equations. The coefficients L_{ii} are called the *primary phenomenological coefficients* and the coefficients L_{ij} the *Onsager phenomenological coefficients*. It should be noted that in the Onsager coefficients the subscript i refers to the flux and the subscript j to the driving force. The primary driving force for J_i is X_i. This has associated with it the primary coefficient L_{ii}. All other coefficients L_{ij} $(i \neq j)$ are also known as *coupling coefficients*.

A solution of equation (12.2) represents a formidable problem because, in general, the coefficients L_{ij} must be determined experimentally. Measurements of the primary coefficients are not very difficult to perform, but experimental determinations of coupling coefficients can be very tedious because of the larger number of parameters that must be kept under strict control. An analytical relationship which helps to alleviate some of the difficulty has been proposed by Onsager. This is now known as Onsager's reciprocal relation. It states that

$$L_{ij} = L_{ji} \tag{12.3}$$

This relation applies to systems in the absence of a magnetic field. A proof of the Onsager relation must be undertaken with statistical thermodynamics. In a physical sense, this relation indicates that coupled flows are induced in a reciprocal manner between interacting driving forces. Equation (12.3) is valid so long as the flows and forces appearing in the phenomenological equations are taken in such a way that

$$\sigma = J_1 X_1 + J_2 X_2 + \ldots + J_n X_n \tag{12.4}$$

where σ represents the local entropy production per unit time per unit volume which results from all the irreversible processes taking place in the system. This condition illustrates clearly the importance of evaluating the explicit expression for the entropy production of the system. To give further emphasis to this very important point again, it is repeated that the reciprocal relation restricts the forms of the fluxes J_i and driving forces X_i. As will be seen in the subsequent sections, the use of equation (12.3) plays a central role in the thermodynamics of irreversible processes. However, it can only be employed if equation (12.4) is utilized to choose the proper flux and forces for the phenomenological equations.

Without further physical insight, the positive-definite nature of the entropy production allows us to deduce certain relations governing the magnitude of the phenomenological coefficients. Considering the case of two fluxes and two driving forces, the entropy production is

$$\sigma = J_1 X_1 + J_2 X_2 \tag{12.5}$$

Introducing the expressions for J_1 and J_2 from equation (12.1)

$$\sigma = L_{11} X_1^2 + (L_{12} + L_{21}) X_1 X_2 + L_{22} X_2^2 > 0 \tag{12.6}$$

Since either X_1 or X_2 may be made to vanish, the requirements are obtained that

$$L_{11} X_1^2 \geq 0 \quad \text{and} \quad L_{22} X_2^2 \geq 0 \tag{12.7}$$

so that both primary coefficients, L_{11} and L_{22}, must be positive. Further, the quadratic form (equation 12.6) will remain positive-definite only if the determinant

$$\begin{vmatrix} L_{11} & L_{12} \\ L_{21} & L_{22} \end{vmatrix} = L_{11} L_{22} - L_{12} L_{21} \geq 0 \qquad (12.8)$$

This provides the restriction on the possible magnitudes of the coupling coefficients L_{12} and L_{21}. In view of equation (12.3), this condition becomes

$$L_{11} L_{22} \geq L_{12}^2 \qquad (12.9)$$

The above conditions imposed on the phenomenological coefficients are general in nature and hold for any number of fluxes and forces.

12.3 ENTROPY PRODUCTION IN SYSTEMS INVOLVING HEAT AND CURRENT FLOWS

Consider a solid bar of uniform cross-sectional area A. The ends of the bar are exposed to the reservoirs T_1 and T_2 and the side of the bar is insulated. At each location in the bar, the heat flux is given by Fourier's law

$$J_Q = \frac{Q}{A} = -k \frac{dT}{dx}$$

The corresponding entropy flux associated with this at each location is

$$J_S = \frac{J_Q}{T} = -k \frac{1}{T} \frac{dT}{dx}$$

This entropy flux is not conserved, however. For the element of length dx the entropy production is

$$\sigma_Q = \frac{J_{S, x + dx} - J_{S, x}}{dx}$$

$$= J_Q \left\{ \left(\frac{1}{T} \right)_{x+dx} - \left(\frac{1}{T} \right)_x \right\} \bigg/ dx$$

$$= J_Q \frac{d \left(\frac{1}{T} \right)}{dx} = -J_Q \frac{1}{T^2} \frac{dT}{dx} \qquad (12.10)$$

Similarly, one may derive the entropy production in a bar resulting from conduction of an electric current. In this instance, heat transfer is allowed to the surroundings in the amount of electric energy dissipated. The electric current in the bar is by Ohm's law

$$J_e = \frac{I}{A} = - \lambda \frac{dE}{dx}$$

At each section of the conductor, the electric energy dissipated is equal to the heat transfer to the surroundings, i.e.

$$\delta Q = I \frac{dE}{dx} dx$$

The resulting entropy production is therefore

$$\sigma_e = \frac{\delta Q}{AT} = - \frac{J_e}{T} \frac{dE}{dx} \tag{12.11}$$

In a bar where both electric and thermal gradients are present, the total entropy production is given by the sum of equations (12.10) and (12.11), i.e.

$$\sigma_{total} = - J_Q \frac{1}{T^2} \frac{dT}{dx} - J_e \frac{1}{T} \frac{dE}{dx} \tag{12.12}$$

If equation (12.4) is to be satisfied for a system involving electric and heat conduction, it is apparent from equation (12.12) that the selection of fluxes and driving forces must correspond to the following scheme.

$$J_Q; \quad X_Q = - \frac{1}{T^2} \frac{dT}{dx}$$
$$J_e; \quad X_e = - \frac{1}{T} \frac{dE}{dx} \tag{12.13}$$

The foregoing results may now be applied to the analysis of the physical phenomena associated with combined electric and heat conduction problems. This will be done in the following section.

12.4 APPLICATION TO THERMOELECTRIC CIRCUITS

Writing the heat flux and electric current in phenomenological form according to equation (12.2).

$$J_Q = - L_{QQ} \frac{1}{T^2} \frac{dT}{dx} - L_{Qe} \frac{1}{T} \frac{dE}{dx} \tag{12.14}$$

and

$$J_e = - L_{eQ} \frac{1}{T^2} \frac{dT}{dx} - L_{ee} \frac{1}{T} \frac{dE}{dx} \tag{12.15}$$

Note that for only primary thermal or electric conduction these equations reduce to

$$J_Q = - L_{QQ} \frac{1}{T^2} \frac{dT}{dx} = - k \frac{dT}{dx} \tag{12.16}$$

and

$$J_e = - L_{ee} \frac{1}{T} \frac{dE}{dx} = - \lambda \frac{dE}{dx} \tag{12.17}$$

whence, the conductivities may be expressed in terms of the primary coefficients by

$$k = \frac{L_{QQ}}{T^2} \quad \text{or} \quad L_{QQ} = kT^2 \tag{12.18}$$

and

$$\lambda = \frac{L_{ee}}{T} \quad \text{or} \quad L_{ee} = \lambda T \tag{12.19}$$

Consider now a coupled one-dimensional flow of electricity and heat. Consider first the circumstances under which there is no current in the material, i.e. when $J_e = 0$. Then from equation (12.15)

$$\left(\frac{dE/dx}{dT/dx} \right)_{J_e = 0} = \left(\frac{dE}{dT} \right)_{J_e = 0} = - \frac{L_{eQ}}{T L_{ee}} \tag{12.20}$$

Equation (12.20) expresses the fact that even with zero current flow, an electric potential will exist when a temperature gradient is present. This phenomenon is called the *Seebeck effect* and the gradient is called the Seebeck coefficient or the thermoelectric power of the material. Defining

$$\alpha = - \left(\frac{dE}{dT} \right)_{J_e = 0} \tag{12.21}$$

and combining equations (12.20) and (12.21), yields

$$L_{eQ} = \alpha T L_{ee} = \alpha \lambda T^2 \tag{12.22}$$

From equations (12.14) and (12.15) with $J_e = 0$, the heat flux is given by

$$J_Q = \frac{-L_{ee} L_{QQ} + L_{eQ} L_{Qe}}{L_{ee} T^2} \frac{dT}{dx} \tag{12.23}$$

Now the thermal conductivity of the coupled system is defined by

$$k = - \frac{J_Q}{\frac{dT}{dx}} \Bigg|_{J_e = 0}$$

hence

$$k = \frac{L_{ee} L_{QQ} - L_{Qe} L_{eQ}}{L_{ee} T^2} \qquad (12.24)$$

Combining with equation (12.22), this gives

$$k = \frac{L_{QQ} - \alpha T L_{Qe}}{T^2}$$

By Onsager's relation

$$L_{Qe} = L_{eQ} = \alpha \lambda T^2$$

therefore

$$L_{QQ} = kT^2 + \alpha T (\alpha \lambda T^2)$$

$$= kT^2 + \lambda \alpha^2 T^3$$

or

$$L_{QQ} = T^2 (k + \lambda \alpha^2 T) \qquad (12.25)$$

Defining the electric conductivity as the electric flux per unit potential gradient under isothermal conditions, the same relation as equation (12.19) is obtained for L_{ee}. Hence, for the case of $J_e \neq 0$

$$J_Q = - (k + \lambda\alpha^2 T) \frac{dT}{dx} - \alpha\lambda T \frac{dE}{dx} \qquad (12.26)$$

$$J_e = - \alpha\lambda \frac{dT}{dx} - \lambda \frac{dE}{dx} \qquad (12.27)$$

By eliminating dE/dx and dT/dx from the above equations, the respective expressions for the total heat and electric fluxes due to the coupled flow are obtained as follows

$$J_Q = - k \frac{dT}{dx} + \alpha T J_e \qquad (12.28)$$

and

$$J_e = - \frac{\lambda k}{k + \lambda \alpha^2 T} \frac{dE}{dx} + \alpha \lambda J_Q \qquad (12.29)$$

From these equations, the effect of coupling on the primary flows may be estimated.

Now consider the junction of two dissimilar metals a and b as shown in Figure 12.1. A current flux J_e is forced through the junction and equilibrium is established at isothermal conditions where the electric energy is dissipated to the surroundings. For isothermal conditions

$$J_e = - \lambda \frac{dE}{dx} \qquad (12.30)$$

254

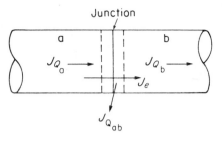

Figure 12.1

In accordance with equation (12.28), the heat flux through each conductor is

$$J_{Q_a} = \alpha_a T J_e \tag{12.31}$$

$$\left. \begin{array}{c} \end{array} \right\} \quad \text{since} \quad \frac{dT}{dx} = 0$$

and

$$J_{Q_b} = \alpha_b T J_e \tag{12.32}$$

The heat interaction at the junction with the surroundings is therefore

$$J_{Q_{ab}} = J_{Q_a} - J_{Q_b} = T(\alpha_a - \alpha_b) J_e = \pi_{ab} J_e$$

The foregoing phenomenon is called the Peltier effect and π_{ab} the Peltier coefficient. It is to be noted that the heat flux $J_{Q_{ab}}$ is due solely to the junction of the two dissimilar metals and *is not* the result of the conventional I^2R or joulean heating. From the above relation, it can readily be deduced that the Peltier coefficient is related to the Seebeck coefficients of the metals forming the junction as follows

$$\pi_{ab} = T(\alpha_a - \alpha_b) \tag{12.33}$$

Equation (12.33) has been known as the second Kelvin relation.

Finally, consider a homogeneous conductor along which a heat flux and an electric flux are flowing and the heat generated is dissipated to the surroundings. The total energy flux J_E through the conductor is given by

$$J_E = J_Q + J_e E$$

which can be combined with equation (12.28) to yield

$$J_E = -k\frac{dT}{dx} + (\alpha T + E) J_e \tag{12.34}$$

The heat interaction with the surroundings that results from a gradient in this total energy flux is

$$dJ_{Q,\text{sur}} = J_{E,x+dx} - J_{E,x} = \frac{dJ_E}{dx} dx \tag{12.35}$$

Hence

$$dJ_{Q,\text{sur}} = \frac{d}{dx}\left[- k \frac{dT}{dx} + (\alpha T + E) J_e \right] dx \qquad (12.36)$$

Assuming that J_e is constant throughout the conductor, equation (12.36) can be expressed as

$$\frac{dJ_{Q,\text{sur}}}{dx} = - k \frac{d^2 T}{dx^2} - \frac{dT}{dx}\frac{dk}{dx} + J_e \left(T \frac{d\alpha}{dx} + \alpha \frac{dT}{dx} + \frac{dE}{dx} \right)$$

Using equations (12.27), the last two terms of the above equation can be written as $- J_e/\lambda$; and if it is further assumed that the bar has a constant temperature gradient, this becomes, since by our assumption of homogeneity k is not a function of position,

$$\frac{dJ_{Q,\text{sur}}}{dx} = T J_e \frac{d\alpha}{dx} - \frac{J_e^2}{\lambda} \qquad (12.37)$$

Equation (12.37) indicates that the heat interaction with the surroundings for this special case is not just equal to the joulean heat production alone. The additional term represents the reversible heating or cooling effect (depending on the direction of J_e) experienced by the conductor because the current is flowing through a temperature gradient. This effect is called the Kelvin heat, which is defined in terms of the Kelvin coefficient by the following expression

$$J_{Q,K} = \gamma J_e \frac{dT}{dx}$$

whence, on substituting the first term of equation (12.37) for $J_{Q,K}$

$$\gamma = T \frac{d\alpha}{dT} \qquad (12.38)$$

This result is known as the first Kelvin relation, which shows that the Kelvin effect arises only when the Seebeck coefficient is a function of temperature. Note also that the minus sign for the Joulean heat term in equation (12.37) is in agreement with the thermodynamic convention for the heat interaction.

Now, for a junction of dissimilar metals, equation (12.38) can be written as

$$\gamma_a - \gamma_b = T \frac{d(\alpha_a - \alpha_b)}{dT} \qquad (12.39)$$

which on combining with the second Kelvin relation yields

$$\frac{d\pi_{ab}}{dT} = (\alpha_a - \alpha_b) + (\gamma_a - \gamma_b) \qquad (12.40)$$

Equation (12.40) provides the interrelation among the Peltier, Seebeck and Kelvin coefficients.

12.5 DIRECT ENERGY CONVERSION FROM THERMOELECTRIC EFFECTS

To round off the presentation, the use of the thermoelectric effect for the production of power and refrigeration will be discussed. Since this mode of power generation has come to be known as direct energy conversion (or in short, DEC) we shall, for the sake of completeness, also attempt to introduce the basic concepts concerning DEC systems, though these generally do not come under the domain of irreversible thermodynamics.

By direct energy conversion we mean the conversion of chemical, nuclear, solar and other forms of thermal energies into electrical energy without the use of mechanical elements such as rotary or reciprocating machineries. With this definition, DEC devices are incapable of providing much better utilization of the world's fuel resources. This is because in most DEC schemes, low grade energy in the form of thermal energy is the source and all such schemes are, like their mechanical counterpart, inherently limited by the Carnot efficiency. Secondly, the utilization of solar energy, the only extraterrestial source of energy, is not particular to DEC.

Then WHY DEC? The answer is that DEC broadens the spectrum of devices for power generation and refrigeration by satisfying certain special performance requirements. Thus, DEC offers a wider choice for specific service, not better devices in the overall sense. In this overall picture, DEC devices fall within the following categories:

(a) cases in which DEC devices open up new application possibilities;
(b) cases in which DEC devices excel over conventional power systems;
(c) cases in which DEC devices provide an alternative;
(d) cases in which DEC devices will not be suitable.

Thus, the choice of DEC systems or conventional power units for any given situation has to be carefully considered.

Characteristic of DEC systems

All DEC devices involve the transport of electrons over potential fields. This transport process is made possible by first exciting the electrons by means of the external source of energy. The flow of these electrons through the external circuit delivers the electrical output. Devices for DEC may be classified as follows.

(1) Direct electron transport. Heat generation is incidental but not essential for their operation. The two general types in this class are fuel cells and radiation (nuclear and solar) cells. Devices in this category, in general, convert energy from a source into the movement of charge carriers. This mode of conversion, therefore, is not limited by the Carnot efficiency.

(2) Energy conversion via heat. Regardless of the source of energy, devices of this type convert heat into electrical energy by thermoelectric or

thermionic elements. (The Magnetohydrodynamic generator may be thought of as belonging to this second class.)

Thermoelectricity for power generation

There are several possible configurations in which thermoelectric materials can be arranged to produce electricity. The simplest and most idealized circuit, similar in shape to the common thermocouple, will be considered here. This is shown in Figure 12.2.

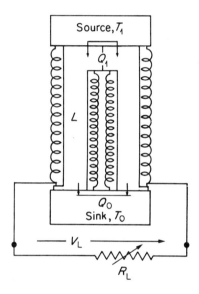

Figure 12.2

In the discussion below, the following assumptions are made about this circuit.

(1) Any variation of the material properties with temperature change is negligible. This implies that the Kelvin heat is zero.
(2) The electrical resistance at the junction interface between the thermoelectric elements is negligible.
(3) The thermoelectric elements have constant areas.
(4) There is no heat loss from the thermoelectric elements to the surroundings.

Based on the above assumptions, the equation of heat conduction for a one-dimensional element with internal joulean heat generation is

$$k \frac{\mathrm{d}^2 T}{\mathrm{d}x^2} = -\rho J^2$$

where k is the thermal conductivity, ρ is the electrical resistivity, J is the current density, and ρJ^2 is the joulean heat per unit volume.

On nondimensionalizing with $\Delta T = T_1 - T_0$ and L

$$\frac{d^2\xi}{d\eta^2} = -\frac{\rho J^2 L^2}{k\Delta T} = -\frac{\rho I^2 L/A}{kA\Delta T/L} = -\frac{Q_J}{Q_F} = -\beta \qquad (12.41)$$

where

$$\xi = (T - T_0)/(T_1 - T_0)$$

$$\eta = x/L$$

Q_J = total joulean heat generated in the bar (since $\rho L/A$ = total resistance of the bar)

Q_F = fourier heat of conduction through the bar = $kA\Delta T/L$

The boundary conditions are

$$x = 0, \quad T = T_1 \qquad \qquad \eta = 0, \quad \xi = 1$$

$$\text{or}$$

$$x = L, \quad T = T_0 \qquad \qquad \eta = 1, \quad \xi = 0$$

On integrating, equation (12.41) becomes

$$\xi = \frac{1}{2}\beta(\eta - \eta^2) + (1 - \eta)$$

The heat flow at end $x = 0$ is

$$Q_1 = -kA\frac{dT}{dx}\Big|_{x=0} = -\frac{kA\Delta T}{L}\left(\frac{\beta}{2} - 1\right) = Q_F - \frac{Q_J}{2} \qquad (12.42)$$

Thus the net heat flow into the bar from the heat source equals the fourier heat less one half of the joulean heat generated internally within the bar. Applying the above result to the two legs of the thermoelectric generator and also taking the peltier heat into account, the heat input into the thermoelectric system can be written as

$$Q_1 = Q_P + Q_F - \tfrac{1}{2}Q_J \qquad (12.43)$$

where the individual terms on the right hand side of equation (12.43) can be written as

$$Q_P = \pi_{pn} I = \alpha_{pn} I T_1$$

$$Q_F = (K_p + K_n)(T_1 - T_0) = \left\{\left(\frac{kA}{L}\right)_p + \left(\frac{kA}{L}\right)_n\right\}(T_1 - T_0)$$

$$Q_J = (R_p + R_n)I^2 = \left\{\left(\frac{\rho L}{A}\right)_p + \left(\frac{\rho L}{A}\right)_n\right\}I^2$$

On substituting into equation (12.43),

$$Q_1 = \alpha_{pn} I T_1 - \tfrac{1}{2}(R_p + R_n)I^2 + (K_p + K_n)(T_1 - T_o) \qquad (12.44)$$

The useful power generated, or work output, in the external circuit is

$$P = R_L I^2 = V_L^2/R_L \qquad (12.45)$$

where

$$V_L = \alpha_{pn}(T_1 - T_o) - (R_p + R_n)I$$

Now, by Kirchhoff's law on electric circuits

$$I = \frac{\alpha_{pn}(T_1 - T_o)}{R_p + R_n + R_L} \qquad (12.46)$$

Let $m = R_L/(R_p + R_n)$, then equation (12.46) can be written as

$$I = \frac{\alpha_{pn}(T_1 - T_o)}{(1 + m)(R_p + R_n)}$$

On substituting for I, equations (12.45) and (12.44) become

$$P = \frac{m}{(1 + m)^2}\frac{\alpha_{pn}^2(T_1 - T_o)^2}{R_p + R_n} \qquad (12.47)$$

and

$$Q_1 = \alpha_{pn}^2 \frac{T_1(T_1 - T_o)}{(R_p + R_n)(1 + m)} - \frac{1}{2}\frac{\alpha_{pn}^2(T_1 - T_o)^2}{(1 + m^2)(R_p + R_n)}$$

$$+ (K_p + K_n)(T_1 - T_o) \qquad (12.48)$$

Therefore, the thermal efficiency of the thermoelectric generator is

$$\eta = \frac{P}{Q_1} =$$

$$\frac{\dfrac{m}{(1+m)^2}\dfrac{\alpha_{pn}^2}{(R_p + R_n)}(T_1 - T_o)^2}{\alpha_{pn}^2\dfrac{T_1(T_1 - T_o)}{(R_p + R_n)(1 + m)} - \dfrac{1}{2}\dfrac{\alpha_{pn}^2(T_1 - T_o)^2}{(1 + m)^2(R_p + R_n)} + (K_p + K_n)(T_1 - T_o)}$$

which can be rewritten as

$$\eta = \frac{T_1 - T_o}{T_1} \frac{m}{(1 + m) - \frac{T_1 - T_o}{2T_1} + \frac{(1 + m)^2 (K_p + K_n)(R_p + R_n)}{\alpha_{pn}^2 T_1}} \quad (12.49)$$

Thus, the thermal efficiency of a thermoelectric system is seen to be a product of two terms. One of these is the Carnot efficiency and the other a term involving all the characteristics of the thermoelectric circuit, i.e. its load ratio, material properties, geometry and operating temperatures. Since the thermoelectric generator is operating between two heat reservoirs, it is a heat engine, therefore the second term, which we shall refer to as the material efficiency of the thermo-electric generator, cannot be greater than unity. As the material efficiency consists of several parameters of the thermoelectric system, the possibility of seeking the optimum design values for these parameters exists. Ignoring for the time being the load ratio m, we see that the material efficiency becomes maximum when the product $(K_p + K_n) (R_p + R_n)$ has its minimum value. Denoting this product by KR

$$KR = \left\{ \left(\frac{kA}{L}\right)_p + \left(\frac{kA}{L}\right)_n \right\} \left\{ \left(\frac{\rho L}{A}\right)_p + \left(\frac{\rho L}{A}\right)_n \right\}$$

$$= (k_p + rk_n) \left(\frac{\rho_n}{r} + \rho_p \right)$$

where

$$r = \left(\frac{A}{L}\right)_n \bigg/ \left(\frac{A}{L}\right)_p$$

represents the ratio of the characteristic lengths of the thermoelectric elements.
On optimizing this

$$\frac{d(KR)}{dr} = k_n \left(\frac{\rho_n}{r} + \rho_p \right) - \frac{\rho_n}{r^2} (k_p + rk_n) = 0$$

or

$$k_n \rho_p - \frac{\rho_n k_p}{r^2} = 0$$

Since $d^2(KR)/dr^2 > 0$, we indeed have a minimum value for KR, whence

$$r_{opt} = \sqrt{\frac{\rho_n k_p}{\rho_p k_n}} \quad (12.50)$$

Thus

$$KR_{opt} = \{\sqrt{(\rho_n k_n)} + \sqrt{(\rho_p k_p)}\}^2 \qquad (12.51)$$

Hence equation (12.49) becomes

$$\eta = \eta_{Carnot} \frac{m}{(1+m) - \dfrac{T_1 - T_o}{2T_1} + \dfrac{\{\sqrt{(\rho_n k_n)} + \sqrt{(\rho_p k_p)}\}^2}{\alpha_{pn}^2} \dfrac{(1+m)^2}{T_1}}$$

The parameter $[\alpha_{pn}/\{\sqrt{(\rho_n k_n)} + \sqrt{(\rho_p k_p)}\}]^2$ which consists only of material properties of the thermoelectric elements, has been referred to as the *figure of merit Z* of the thermoelectric elements. On making this substitution, the above expression becomes

$$\eta = \eta_{Carnot} \frac{m}{(1+m) - \dfrac{T_1 - T_o}{2T_1} + \dfrac{(1+m)^2}{ZT_1}} \qquad (12.52)$$

Further optimization of η can now be continued by optimizing with respect to m, i.e. by choosing the load ratio such that the material efficiency will attain a maximum value. On differentiating with respect to m and equating to zero

$$(1+m) - \frac{T_1 - T_o}{2T_1} + \frac{(1+m)^2}{ZT_1} - m\left\{1 + \frac{2(1+m)}{ZT_1}\right\} = 0 \qquad (12.53)$$

which can be solved to give

$$m_{opt} = \sqrt{\left\{1 + \frac{Z(T_1 + T_o)}{2}\right\}} \qquad (12.54)$$

Using equations (12.53) and (12.54), the optimum thermal efficiency can be deduced from equation (12.52) as

$$\eta_{opt} = \eta_{Carnot} \frac{m_{opt} - 1}{m_{opt} + \dfrac{T_o}{T_1}} \qquad (12.55)$$

It is therefore seen that the thermoelectric device can never achieve the efficiency allowed by the second law of thermodynamics, since this can only be attained as m_{opt} (and hence Z) tends to infinity. Thus, for systems at optimum design, Z is the only variable left to be selected; thus a pair of thermoelectric elements should give as large a value for Z as possible. Once this has been chosen, it then determines the optimum external load at which maximum efficiency can be attained.

For certain applications, one may wish to have a maximum power output by sacrificing the optimum economic operation at maximum efficiency. The efficiency at maximum power can be obtained by optimizing equation (12.47) with respect to m and then substituting this value for m in equation (12.52). It is easy to show that the condition for maximum power corresponds to $m = 1$, whence

$$\eta_{\text{max power}} = \frac{\eta_{\text{Carnot}}}{2 - \frac{1}{2}\eta_{\text{Carnot}} + \frac{4}{ZT_1}} \tag{12.56}$$

Properties of thermoelectric materials

It has been demonstrated in the previous section that the contribution of material properties to the optimum efficiency of the thermoelectric generator can be summarized in a single factor known as the figure of merit of the thermoelectric elements. It will briefly be shown how this factor is influenced by the basic material properties of a given material. For a single material, the definition for Z reduces to

$$Z = \frac{\alpha^2}{\rho k} = \frac{\alpha^2 \sigma}{k}$$

where $\sigma = 1/\rho$ is known as the electrical conductivity of the material. All the three basic properties involved in the above definition of Z are to some extent measures of the concentration as well as the mobility of the charge carriers in the material. Figure 12.3(a) shows how each of these properties varies with the carrier electron concentration, and Figure 12.3(b) and (c) display the successive combination of the basic properties leading to the variation of Z with respect to carrier electron concentration. From thèse figures, it is easy to see why semiconductors are preferred for use as thermoelectric elements. The advancement of thermoelectric systems, therefore, hinges to a great extent on our ability to develop suitable materials.

Outlook for thermoelectric power generation

The thermoelectric power generator enjoys several advantages over conventional power plants. These include quietness of operation, freedom from vibration, low maintenance costs, relatively simple controls and an efficiency independent of its power rating. (Gas cycle engines become less efficient as their size decreases.) Thus, thermoelectric generators are expected to find a ready market in the small generator range provided that their capital costs are competitive with other generators available. As far as efficiency is concerned, present thermoelectric generators having conversion efficiencies in the range of 6–10 percent are already competitive with other heat engines with power ratings

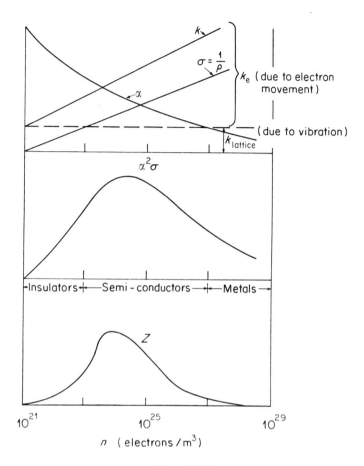

Figure 12.3

below 10 kW. Their costs are, however, high at the present time. This situation may change when systems having conversion efficiencies approaching 20 percent are put on the market. The high price of fuel may then favour systems having much lower operating costs in the long run, as this will eventually be able to offset the initial capital investment. The ultimate theoretical limit for thermo-electric conversion efficiency has been estimated at 36 percent.

On an even smaller scale, the use of thermoelectric devices opens up the possibilities of increased miniaturization and in many cases further simplifica-tion of present day designs.

PROBLEMS

1. The junctions of a thermoelectric generator are kept in two different uniform temperature baths such that a total emf of 0.8 V is produced in the open circuit condition. The leads are connected to an ideal electric motor and the current flow is 100 A.

(a) How much work is done by the generator if the internal resistance of generator arms is 5 mΩ?

(b) If the efficiency of the system is 7 percent, find the heat interactions at the high and low temperature junctions.

(c) Will the heat interactions of (b) be changed if the leads are connected to an outside resistance of 3 mΩ instead of the motor, and if so, by what amount?

(d) For a hot reservoir of 600 K and a cold reservoir of 300 K, find the rate of entropy change of the entire system.

(e) A reversible engine is receiving heat from the same reservoirs as in (d). Find the rate of entropy change. Hence, determine the irreversibility of the thermoelectric generator system.

2. The thermoelectric power generation circuit can be reversed to remove heat Q_o from the reservoir at T_o and to reject Q_1 to the reservoir at T_1 by means of power input P. Show that when used to provide thermoelectric cooling, the coefficient of performance at maximum cooling effect is given by

$$(COP)_{max\ cooling} = \frac{1}{2}\frac{T_o}{T_1} - \frac{T_1 - T_o}{ZT_1 T_o}$$

3. Show that the maximum coefficient of performance of a thermoelectric refrigerator is given by

$$\frac{T_o}{T_1 - T_o}\frac{m_{opt} - \dfrac{T_1}{T_o}}{1 + m_{opt}}$$

where

$$m_{opt} = \sqrt{\left\{1 + \frac{Z}{2}(T_o + T_1)\right\}}$$

Index

268

272